资阳美食文化

主 编 杜莉 陈祖明
Editors-in-Chief　Du Li, Chen Zuming

ZIYANG
FOOD CULTURE

SICHUAN PUBLISHING HOUSE OF SCIENCE & TECHNOLOGY　四川科学技术出版社

图书在版编目（CIP）数据

资阳美食文化/杜莉，陈祖明主编. -- 成都：四川科学技术出版社，2019.8（2022年4月重印）
ISBN 978-7-5364-9553-1

Ⅰ.①资… Ⅱ.①杜… ②陈… Ⅲ.①饮食—文化—资阳 Ⅳ.①TS971.202.713

中国版本图书馆CIP数据核字(2019)第168881号

资阳美食文化
ZIYANG MEISHI WENHUA

主　编　杜莉　陈祖明

出 品 人	程佳月
责任编辑	程蓉伟
出版发行	四川科学技术出版社
装帧设计	程蓉伟
封面设计	程蓉伟
责任印制	欧晓春
制　　作	成都华林美术设计有限公司
印　　刷	成都市金雅迪彩色印刷有限公司
成品尺寸	285mm×210mm
印　　张	19
字　　数	300千
版　　次	2019年9月第1版
印　　次	2022年4月第2次印刷
书　　号	ISBN 978-7-5364-9553-1
定　　价	258.00元

■ 版权所有·侵权必究
　本书若出现印装质量问题，联系电话：028-87733982

资

阳

滋

味

《资阳美食文化》编纂委员会

顾 问
廖仁松　晏启鹏　吴旭　卢一　王荣木　陈莉萍

主 任
周密　许志勋　陈云川

副主任
孙智　邓爱华　余贵冰　杜莉　周世中　陈祖明

委 员
（以姓氏笔画为序）

王胜鹏　冉伟　朱健　刘军丽　刘官银　李力　肖述明　吴先锋　吴茂森
张达　张茜　张媛　陈筱　罗刚　罗霄　郑伟　秦照明　聂瑞刚　黄洁玉　程蓉伟

主 编
杜莉　陈祖明

副主编
王胜鹏　刘军丽　郑伟

中文撰稿及审校
杜莉　陈祖明　王胜鹏　刘军丽　郑伟　张茜　贺树

英文翻译及审校
张媛　张琪　张涪涪　李潇潇　彭施龙　王畅

图片拍摄及菜点造型
程蓉伟　冉伟　陈筱

协助拍摄
资阳市餐饮协会　方圆宾馆　小贝壳酒店　美食美味大酒店　柠檬花精品酒店

编纂单位
资阳市商务局　四川旅游学院

Compilation Committee of *Ziyang Food Culture*

Counselors
Liao Rensong, Yan Qipeng, Wu Xu, Lu Yi, Wang Rongmu, Chen Liping

Directors
Zhou Mi, Xu Zhixun, Chen Yunchuan

Deputy Directors
Sun Zhi, Deng Aihua, Yu Guibing, Du Li, Zhou Shizhong, Chen Zuming

Committee Members
Wang Shengpeng, Ran Wei, Zhu Jian, Liu Junli, Liu Guanyin, Li Li, Xiao Shuming, Wu Xianfeng, Wu Maosen, Zhang Da, Zhang Qian, Zhang Yuan, Chen Xiao, Luo Gang, Luo Xiao, Zheng Wei, Qin Zhaoming, Nie Ruigang, Huang Jieyu, Cheng Rongwei

Editors-in-Chief
Du Li, Chen Zuming

Deputy Editors-in-Chief
Wang Shengpeng, Liu Junli, Zheng Wei

Chinese Compiled and Reviewed by
Du Li, Chen Zuming, Wang Shengpeng, Liu Junli, Zheng Wei, Zhang Qian, He Shu

English Translated and Reviewed by
Zhang Yuan, Zhang Qi, Zhang Tiantian, Li Xiaoxiao, Peng Shilong, Wang Chang

Photography and Dishes Designed by
Cheng Rongwei, Ran Wei, Chen Xiao

Photography Assisted by
Ziyang Catering Association, Fangyuan Hotel, Little Shell Hotel, Gourmet Restaurant, Lemon Flower Boutique Hotel

Compilation Units
Ziyang Municipal Bureau of Commerce,
Sichuan Tourism University

资阳雁江区三贤公园
The Three Sages Park in Yanjiang District of Ziyang

序

今年是我大学入校四十周年。1979年9月初，我怀揣四川医学院（曾更名"华西医科大学"，现已并入四川大学）的录取通知书，取道宜宾，坐上火车前往成都。当火车途经资阳时，看见人们纷纷在月台上购买一种叫"临江寺豆瓣"的资阳特产，这是那个年代资阳火车站的一道独特风景。细细地想来，我对资阳的记忆是从美食开始的。

我从华西医科大学硕士研究生毕业后，来到四川烹饪高等专科学校（现升格为"四川旅游学院"）从事烹饪教学和研究工作，特别专注于川菜和饮食文化的研究。有一次，听说资阳孔庙内的孔圣人的塑像造型与其他地方的不一样，因为这里有孔子的老师苌弘。原来资阳是苌弘的故乡，孔子曾专程造访、拜苌弘为师，请教音乐理论和天文知识。此后，孔子便写出了《乐经》这一不朽著作。该书与《诗经》《尚书》《礼经》《易经》《春秋》一起，共同成为了儒家经典中最著名的"六经"。西汉时期，同样为资阳人的著名文学家王褒在《僮约》中两次提到茶，无意间成为最早记载中国茶文化的文献，为中国茶史留下了非常宝贵的一笔。其中，"脍鱼炮鳖，烹茶尽具""武阳买茶"更是研究茶文化、中国茶史的人们经常引用的语句。这由此说明，在西汉时的四川，饮茶习俗颇为盛行，不仅有茶具，而且茶叶已成为商品，有了交易市场，武阳（今眉山市彭山区江口镇）便是其中的一个茶叶交易市场。从苌弘和王褒的事迹便可见资阳文化底蕴深厚，文脉传承清晰，成为资阳美食传承发展的重要源泉和根基。

2019年秋天，世界中餐业联合会与资阳市人民政府联合主办"第二届世界川菜大会"，四川旅游学院参与承办。此次大会以"让世界爱上川菜""爱上川菜 爱上资阳"为主题，将举办"2019全球川菜产业发展高峰论坛""川菜'走出去'城市发展论坛""2019世界四川火锅宴大赛"等多个子活动。期间，由四川旅游学院杜莉教授、陈祖明教授率领项目团队主编的《资阳美食文化》一书将与国内外读者和社会各界见面。该书在广泛搜集和整理相关资料、进行大量调研的基础上，深入挖掘资阳美食文化涉及的各个方面，不仅全面、深入地研究和论述了资阳美食的历史沿革与发展

现状，归纳、总结出资阳美食特色，而且系统地筛选、阐述了资阳名特食材、名人与美食典故、名茶与名酒以及美食街区、美食小镇、美食名店的状况，系统地筛选、记载了资阳市代表性菜肴与面点小吃的食材配方、制法和风味特征。这无疑是一件十分重要和有意义的事。通过上述系列活动和该书的出版，将进一步搭建全球餐饮产业交流平台，加强川菜与各区域特色餐饮集群互动，推动川菜与世界对话，同时也将进一步推动资阳美食文化的传承和创新，特别是推动餐饮业供给侧结构性改革，促进资阳当地特色农产品（食材）的品牌化、产业化和国际化发展，扩大内需，既能满足人们对美好生活的向往，也能增加农民收入，加快乡村振兴、全面建成小康社会的步伐。

文化只有在创新和发展中才能更好地传承，否则，随着经济发展和技术进步，很多传统技艺将因劳动力成本增高而无力传播和推广，面临失传的危险，而仅仅依靠"非遗"保护也较为有限。因此，生产更多的美食文创产品，实现优秀传统文化的创造性转化、创新性发展，将有利于我们更加坚定文化自信，有利于资阳在文化和美食领域的永续辉煌。

自20世纪80年代以来，资阳行政区划发生了许多变化。如今，随着"成渝门户枢纽，临空新兴城市"的建设，资阳还将迎来更多的发展机遇和空间，生活在这片土地上的人民，他们的文化和生活方式、独特的美食及饮食习俗将在发展变化中永存。

是为序。

卢一

写于成都廊桥南岸小鲜斋

2019年7月30日

PREFACE

This year marks the 40th anniversary of my admission to the university. In early September, 1979, I took the train to Chengdu via Yibin with the admission notice of Sichuan Medical College (once renamed West China University of Medical Sciences, now incorporated into Sichuan University). When the train passed through Ziyang, I saw people on the platform buy a kind of Ziyang specialty called "Linjiangsi chili bean paste". This was a unique view of Ziyang Railway Station at that time. Think about it carefully, my memory of Ziyang began with food.

After I graduated from West China University of Medical Sciences with a master's degree, I came toSichuan Higher Institute of Cuisine (now upgraded to Sichuan Tourism University) to engage in cooking teaching and research, especially the research on Sichuan cuisine and food culture. Once, I heard that the statues of Confucius in ConfucianTemples in Ziyang were different from those in other places, because there was a man named Chang Hong, who was the teacher of Confucius. Ziyang was Chang Hong's hometown. Confucius ever paid a special visit to Chang Hong and took him as the teacher to consult music theory and astronomical knowledge. After that, Confucius wrote the monumental work, *Classic of Music*, which, together with *The Book of Songs*, *The Book of History*, *The Book of Rites*, *The Book of Changes*, and *The Spring and Autumn Annals*, became the most famous "Six Classics" among Confucian classics. In the Western Han Dynasty, Wang Bao, a famous litterateur who was also from Ziyang, mentioned tea twice in his work *Tongyue*, which inadvertently became the earliest document recording Chinese tea culture and left a very precious record in the history of Chinese tea. Among them, "slicing fresh fish and stewing turtle, boiling tea and cleaning tea set", and "buying tea in Wuyang" are often quoted by people who study tea culture and history of Chinese tea. All these show that in the Western Han Dynasty, the custom of drinking tea was very popular in Sichuan. There were not only tea sets, but also tea had become a commodity and had trading markets. Wuyang (now Jiangkou Town, Pengshan District, Meishan City) was one of the tea trading markets. We can see from Chang Hong's and Wang Bao's deeds that Ziyang has profound cultural background and clear cultural heritage, which have become the important source and foundation for the inheritance and development of Ziyang cuisine.

In the autumn of 2019, the World Association of Chinese Cuisine and Ziyang Municipal People's Government will jointly host The 2nd World Sichuan Cuisine Conference, which Sichuan Tourism University will participate in the undertaking. With the themes of "let the world love Sichuan cuisine", and "fall in love with Sichuan cuisine and fall in love with Ziyang", the conference will hold several sub-activities such as 2019 Global Summit Forum on Sichuan Cuisine Industry Development, Sichuan Cuisine "Going Out" Urban Development Forum, and 2019 World Sichuan Hot Pot Banquet Competition. During this period, the book *Ziyang Food Culture*, edited by the project team led by Professor Du Li and Professor Chen Zuming of Sichuan Tourism University, will meet with readers at home and abroad and all walks

of life. Based on extensive collection and collation of relevant materials andmass investigation and research, the book digs deeply into all aspects involved in Ziyang food culture. This book not only comprehensively and further studies and discusses the history and development of Ziyang cuisine, and summarizes the characteristics of Ziyang cuisine, but also systematically selects and expounds the famous special food materials, celebrities and allusions, well-known liquors and tea drinks, as well as food streets, towns and restaurants of Ziyang, and documents the ingredients and recipes, preparations and flavor characteristics of the representative dishes and snacks of Ziyang. This is undoubtedly a very important and meaningful thing. Through the above-mentioned series of activities and the publication of the book, we will further build a communicative platform of global catering industry, strengthen the interaction between Sichuan cuisine and regional catering clusters, promote the dialogue between Sichuan cuisine and the world, and meanwhile will further advance the inheritance and innovation of Ziyang food culture, especially push forward the supply-side structural reform of catering industry, facilitate the branding, industrialization and internationalization development of local characteristic agricultural products (food materials) in Ziyang, which can not only satisfy people's yearning for a better life, but also increase farmers' income, accelerate the pace of rural revitalization and building a moderately well-off society in all aspects.

Only in innovation and development can culture be better inherited. Otherwise, with economic development and technological progress, many traditional skills can not be disseminated and promoted and in danger of being lost because of the increased labor costs. Only relying on "intangible cultural heritage" protection is also relatively limited. Therefore, the production of more catering cultural and creative products and the realization of creative transformation and innovative development of excellent traditional culture will help us to strengthen cultural self-confidence and Ziyang's sustainable brilliance in the fields of culture and food.

Since the 1980s, many changes have taken place in the administrative division of Ziyang. Nowadays, with the construction of "Chengdu-Chongqing Portal Hub, Airport Emerging City", Ziyang will usher in more opportunities and space for development. People living in this land, their culture and lifestyle, unique food and dietary customs will survive in the development and change.

That is what I want to say for the book.

Lu Yi
Written in Xiaoxianzhai on the
south bank of Langqiao Bridge in
Chengdu July 30th, 2019

前言

　　资阳，地处天府之国的中西部、沱江西岸，现总辖雁江区、安岳县和乐至县，总面积约5 757平方公里，是同时连接成渝"双核"的四川省区域性中心城市。资阳美食依托丰富的自然资源、悠久的历史，在政府引导、协会助推、市场主导、企业运作、立足民间等原则的指导下，造就了特色鲜明、异彩纷呈的美食文化，初步形成了美食产业链体系，美食产业繁荣兴旺，人才队伍建设与品牌打造也取得显著成效。资阳美食成为了新时代资阳市活色生香、有滋有味的独特城市名片。

　　资阳境内气候温润，地形多样，河流纵横，自古便是四川盆地农业发达的重要区域之一，出产丰富的特色优质食材。这一切滋养着一代又一代资阳人在此生存、繁衍，创造了多姿多彩的资阳美食文化。

　　资阳美食文化底蕴深厚，内涵丰富，涉及众多方面，包括历史发展、食材、菜点及其制作技艺、美食街（镇）与餐饮店、名人与美食典故、茶酒，等等。早在35 000年前，古老的"资阳人"就在这里繁衍生息，创造了沱江流域的人类文明史。资阳美食自史前时代开始，历经夏商周、秦汉魏晋、宋元明清，延绵不断，在各个方面都取得了较大成就。进入20世纪80年代中期，尤其是21世纪以后，在各级政府的支持和推动、行业协会与餐饮从业人员的共同努力下，资阳美食逐渐走向高速发展阶段，呈现出勃勃生机。资阳各地出产着大量优质特色食材，如安岳红薯、安岳柠檬、周礼粉条、乐至藕粉、乐至黑山羊、坛子肉和伍隍猪、临江寺豆瓣等。许多特色食材是国家地理标志产品和四川乃至全国的知名品牌，远销海内外。它们是资阳乡村经济的重要组成部分和农民脱贫致富的重要资源，也是资阳美食文化的靓丽名片。资阳人民依托丰富的特色食材，巧妙地制作了许多特色鲜明的珍馐佳肴，如苌弘鲶鱼、资阳酸酸鸡、水晶柠檬、柠香脆皮鱼、乐至烤肉、周礼伤心凉粉、倒罐蒸桑叶鸡、桑叶酥等。随着社会发展和人民生活水平的提高，资阳琳琅满目的菜点名宴开始更加追求安全、绿色、营养、时尚，成为资阳文化与旅游融合发展的重要资源，是川菜文化宝库中的一颗璀璨明珠。资阳市委、市政府高度重视并大力支持美食产业的发展和转型升级，努力打造特色美食品牌。一方面，积极培育龙头企业，发展壮大中小企业，不断提高美食产业信息化水平，提升美食产品的品质与服务，使资阳拥有了一批知名的餐饮企业和店铺，如金迪大酒店、小贝壳酒店、美食美味大酒店、无名烧鸡店等。另一方面，积极挖掘美食旅游资源，打造美食旅游品牌，名特美食街区和小镇不断涌现，如雁江三贤餐饮特色街、乐至帅乡大道美食街区、安岳中央时代广场美食街区、安岳柠檬小镇、雁江中和镇、乐至劳动镇等。这些资阳美食的名特街区、小镇以及餐饮企业、店铺已成为人们追寻美食、享受美食的理想去处，成为传播、弘扬资阳美食文化的重要载体。资阳这块土地上还先后走出了众多的圣贤与名人，有孔子之师苌弘、汉代文学家王褒和经学家董钧，有无产阶级革命家、军事家、外交家、诗人陈毅元帅，有著名作家和书画家邵子南、周克芹、谢无量等，还有许多资阳美食的烹饪大师、企业家、文化名人。他们的人生故事里有着与资阳美食相连的精彩片段，也成为资阳美食文化的重要组成部分。此外，资阳还拥有众多的名特优质酒品和茶饮，其中以柠檬、桑叶与桑葚制作的茶酒品种更是别具风味。

2019年9月，"第二届世界川菜大会"将在资阳召开，这是资阳美食产业发展进程中的一个新里程碑。为此，四川旅游学院与资阳市政府实施战略合作，其重要内容之一便是开展《资阳美食文化》一书编撰工作。

本书由四川旅游学院川菜发展研究中心、资阳市商务局共同负责，组建了包括饮食文化、烹饪技术与艺术、英语翻译等方面的专家与学者以及烹饪大师在内的项目团队，于2019年5月底开始工作。团队多次赴资阳各地实地调研，与各级政府相关部门、行业协会、餐饮企业家及烹饪大师和名师等座谈，搜集、整理相关资料，首先设计全书的框架大纲，并在征求多方意见和建议的基础上，对本书的名特食材、菜点、茶酒和名特街（镇）与餐饮店、名人与典故等制定了较为明确的入选标准和原则。据此，编撰出全书的条目及初稿，再次提交学院领导、专家及资阳市领导和各方征求意见和建议，并进一步修改、完善，形成定稿、进行英文翻译。

《资阳美食文化》一书全面、系统地梳理和总结了资阳美食文化的整体情况，内容丰富、图文并茂、中英文对照，集学术性、实用性、趣味性于一体。全书由杜莉设计框架大纲并负责总体统筹，共分为六个篇章。其中，第一篇主要由王胜鹏执笔；第二篇和第四篇主要由郑伟执笔；第三篇由陈祖明负责并主要执笔；第五篇和第六篇主要由刘军丽执笔。最后，由杜莉统稿、审核中文稿，张茜、贺树校正，由张媛、张琪、张滟滟、李潇潇、彭施龙、王畅等翻译成英文并审校。陈筱负责资料汇总与沟通。本书第三篇选录的名特菜点共100道，其中的90道来自资阳市雁江区、安岳县、乐至县的餐饮企业、店铺乃至民间家庭，由相关厨师精心制作，由四川科学技术出版社程蓉伟进行装盘设计并拍摄图片，冉伟和陈筱参与其装盘设计；10道菜点由陈廷龙、吴奇安、刘晓旭、唐路明、杨虎成等5位资阳籍川菜大师精心制作并拍摄图片。此外，资阳市政府各级相关部门、企业等为本书提供了相关资料和图片，并做了组织协调工作。

本书在研究和编撰过程中，不仅得到四川旅游学院领导、资阳市政府领导的高度重视和指导，也得到资阳市商务局领导和四川科学技术出版社领导的大力支持，还得到资阳市各级政府相关部门、餐饮界同仁和出版社编辑的积极配合与协作，为本书的编撰、出版奠定了良好基础。可以说，本书是四川旅游学院与资阳市各级政府相关部门、餐饮界同仁和四川科学技术出版社共同努力的结晶，将进一步传承、弘扬资阳美食文化，提升其软实力、国际知名度和影响力，也将进一步丰富川菜文化内涵，推动川菜产业化和国际化发展。在本书即将出版之际，对给予我们关心、指导、支持与帮助的各位领导、专家、学者、业界同仁表示衷心感谢！也对所有参与此项工作、辛勤付出心血与时间，尤其是付出了周末及暑假的朋友们表示衷心感谢！

资阳美食文化内容十分丰富，但由于开展此项编撰出版工作的时间极其短暂，加之能力所限，书中难免有不足甚至错漏之处，恳请领导、专家学者、业界同仁和广大读者不吝赐教，以便今后修订完善。

编　者

2019年8月于成都

FOREWORD

Ziyang, located in the mid-west of the land of abundance and the west bank of the Tuojiang River, has jurisdiction over Yanjiang District, Anyue County and Lezhi County, with the total area of about 5,757 square kilometers. It is a regional central city in Sichuan Province connecting Chengdu-Chongqing "double core". Relying on the abundant natural resources and a long history, under the direction of government guiding, association helping, market leading, enterprise operation, base on the folk and other development principles, Ziyang cuisine creates the distinctive and colorful food culture and forms a food chain system. The food industry is flourishing, and remarkable achievements have been made in the construction of catering talent team and brand creation. Ziyang cuisine has become the lively and unique name card of Ziyang City in the new era.

Ziyang has a warm and moist climate, diverse terrain and crisscrossrivers. It has been one of the most developed agricultural areas in Sichuan Basin since ancient times, producing rich and high-quality food materials. All these nourish generations of Ziyang people to live and multiply here, creating a colorful food culture.

Ziyang food culture is profound and rich in connotations. It involves many aspects, including historical development, food materials, dishes and snacks and the production techniques, food streets (towns) and restaurants, celebrities and food allusions, tea and wine, and so on. As early as 35,000 years ago, the ancient "Ziyang Man" lived and multiplied here, creating the history of human civilization in Tuojiang River Basin. Since prehistoric times, going through the Xia, Shang, Zhou, Qin, Han, Wei, Jin, Song, Yuan, Ming and Qing dynasties, Ziyang cuisine has developed continuously, making great achievements in all aspects. Since the mid-1980s, especially after the 21st century, Ziyang cuisine has gradually entered the high-speed development stage with the support and promotion of governments at all levels and the joint efforts of catering practitioners, showing vigor and vitality. Ziyang produces a large number of high-quality characteristic food materials, such as Anyue sweet potato, Anyue lemon, Zhouli vermicelli, Lezhi lotus root porridge, Lezhi black goat, Tanzi Pork and Wuhuang pig, Linjiangsi chili bean paste, etc. Many characteristic food materials are national geographic indication products and famous brands in Sichuan and even in China, which are sold at home and abroad. They are the important part of the rural economy of Ziyang, the important resources for farmers to get rid of poverty and become rich, and also the beautiful name card of Ziyang food culture. Relying on rich specialty ingredients, the people of Ziyang ingeniously produced many distinctive delicacies, such as Changhong Catfish, Ziyang Suansuan Chicken, Crystal Lemon, Lemon Flavor Crispy Fish, Lezhi Pork BBQ, Zhouli Spicy Pea Jelly, Chicken Steamed with Mulberry Leaves, Mulberry Leaf Fritters, etc. With the development of society and the improvement of people's living standards, the wide variety of dishes and famous banquets of Ziyang begin to pursue safety, green, nutrition and fashion more, becoming an important

resource for the integrated development of Ziyang culture and tourism, and a bright pearl in the treasure house of Sichuan cuisine culture. Ziyang Municipal Party Committee and Municipal Government attach great importance to and support the development, transformation and upgrading of food industry, making great efforts to build special food brand. On the one hand, the government actively cultivates leading enterprises, develops small and medium-sized enterprises, constantly raises the informationalized level of food industry, and improves the quality of food products and services, thus creating a number of well-known catering enterprises and shops in Ziyang, such as Jindi Hotel, Little Shell Hotel, Gourmet Restaurant, Wuming Braised Chicken Store, etc. On the other hand, the government actively develops food tourism resources and builds food tourism brands, special food streets and towns keeping popping up, such as Yanjiang Sanxian Special Food Street, Lezhi Shuaixiang Avenue Food Street, Anyue Central Times Square Food Street, Anyue Lemon Town, Zhonghe Town of Yanjiang District, Labor Town of Lezhi County, etc. These special food streets, small towns, catering enterprises and shops of Ziyang have become ideal places for people to pursue and enjoy delicious food, and become important carriers to spread and carry forward the food culture of Ziyang. There are also many sages and celebrities coming out of Ziyang, including Chang Hong, the teacher of Confucius, Wang Bao, litterateur of the Han Dynasty, Dong Jun, Confucianist of the Han Dynasty, Marshal Chen Yi, the proletarian revolutionist, militarist and diplomat, and famous writers and calligraphers and painters Shao Zinan, Zhou Keqin, Xie Wuliang etc., as well as many cooking masters of Ziyang cuisine, entrepreneurs, cultural celebrities. There exists wonderful moments in their life stories connecting with Ziyang cuisine, which also become the important part of Ziyang food culture. In addition, Ziyang also has a great amount of famous special high-quality liquors and tea drinks, among which the varieties made from lemon, mulberry leaves and mulberries are of particular flavors.

In September, 2019, "The 2nd World Sichuan Cuisine Conference" will be held in Ziyang, which is a new milestone in the development of the food industry in Ziyang. Therefore, Sichuan Tourism University and Ziyang Municipal Government implement the important content of the strategic cooperation to carry out the compilation of the book *Ziyang Food Culture*.

This book is jointly responsible by Sichuan Cuisine Development Research Center of Sichuan Tourism University and Ziyang Municipal Bureau of Commerce. A team of experts and scholars incatering culture, culinary technology and art, English translation and other fields, as well as cooking masters was formed and got to work at the end of May, 2019. The team went to various places in Ziyang for field research, held discussions with relevant government departments at all levels, industry associations, catering entrepreneurs, cooking masters and other experts, and collected and sorted out relevant information. The framework outline of the book was designed first. On the basis of soliciting opinions and suggestions from various parties, relatively clear inclusion criteria and principles were formulated for the famous special food materials, dishes and snacks, liquors and tea drinks, special streets (towns) and restaurants, celebrities and allusions of the book. Accordingly, the entries and first draft of the book were compiled and submitted to the leaders of the college, experts and leaders of Ziyang City for comments and suggestions, and further amendments and improvements were made to form a final draft and carry out English translation.

The book *Ziyang Food Culture* comprehensively and systematically combs and summarizes the overall situation of Ziyang food culture. It is rich in content, illustrated and bilingual in Chinese and English, integrating academic, practical and interesting. The framework outline of the book is designed by Du Li, who is also in charge of overall planning. It is divided into six chapters. Among them, Chapter One is mainly written by Wang Shengpeng, the second and fourth chapters are mainly written by Zheng Wei, Chapter Three is responsible and mainly written by Chen Zuming, and the fifth and sixth chapters are mainly written by Liu Junli. Finally, the Chinese manuscript was compiled and reviewed by Du Li, and proofread by Zhang Qian and He Shu, and was translated into English and reviewed by Zhang Yuan, Zhang Qi, Zhang Tiantian, Li Xiaoxiao, Peng Shilong and Wang Chang, etc. Chen Xiao was responsible for data collection and communication. A total of 100 famous special dishes and snacks were included in the third chapter of this book, 90 of which were from catering enterprises, shops and even folk families in Yanjiang District, Anyue County and Lezhi County of Ziyang City. The dishes were elaborately made by relevant chefs, designed and photographed by Cheng Rongwei from Sichuan Publishing House of Science and Technology. Ran Wei and Chen Xiao participated in the dish design. And the other 10 dishes were carefully produced and photographed by five Ziyang masters of Sichuan cuisine, namely, Chen Tinglong, Wu Qi'an, Liu Xiaoxu, Tang Luming, Yang Hucheng. Besides, relevant departments at all levels of Ziyang Municipal Government and enterprises provided related materials and pictures for this book, and did the organization and coordination work.

During the process of research and compilation, the book not only received the high attention and guidance from the leaders of Sichuan Tourism University and Ziyang Municipal Government, but also got the strong support from the leaders of Ziyang Municipal Bureau of Commerce and Sichuan Publishing House of Science and Technology, as well as the active cooperation from the relevant government departments at all levels in Ziyang City, colleagues in catering circles and press editors, which have laid a good foundation for the compilation and publication of the book. It can be said that this book is the result of the joint efforts of Sichuan Tourism University, relevant government departments at all levels in Ziyang City, colleagues in catering circles, and Sichuan Publishing House of Science and Technology. It will further inherit and carry forward Ziyang food culture, enhance its soft power, international reputation and influence, and also further enrich the cultural connotation of Sichuan cuisine and promote the industrialization and internationalization of Sichuan cuisine. As this book is about to be published, I would like to express my heartfelt thanks to all the leaders, experts, scholars and colleagues in the industry who have given us care, guidance, support and help. I would also like to express my sincere gratitude to all the friends who have participated in this work, and devoted their efforts and time, especially on weekends and summer holidays to the book.

Ziyang food culture is very rich in content. However, due to the extremely short time to compile and publish this book and the limited ability, there are inevitably some shortcomings or even mistakes in the book. Therefore, we sincerely hope that leaders, experts and scholars, colleagues in the industry and the readers can give us suggestions for future revision and improvement.

From the Editor
Written in Chengdu
August, 2019

资阳雁江区沱江三桥
The Third Tuojiang River Bridge
in Yanjiang District of Ziyang City

目 录

001 第一篇 资阳美食的历史沿革与发展状况
Chapter One The History and Development of Ziyang Cuisine

003 壹 资阳美食的历史沿革
History of Ziyang Cuisine

017 贰 当代资阳美食发展现状
Current Situation of the Development of Contemporary Ziyang Cuisine

031 第二篇 资阳美食地理环境与名特食材
Chapter Two Geographical Environment and Famous and Special Food Materials of Ziyang Cuisine

033 壹 地理环境
Geographical Environment

037 贰 名特食材
Famous and Special Food Materials

059 第三篇 资阳名特菜肴与面点小吃
Chapter Three Renowned Dishes and Snacks of Ziyang

060 壹 冷菜
Cold Dishes

074 贰 热菜
Hot Dishes

190 叁 面点小吃
Snacks

CONTENTS

213 第四篇 资阳名特美食街（镇）与餐饮店
Chapter Four   Ziyang Famous Special Food Streets (Towns) and Restaurants

215 壹 名特美食街区与小镇
Famous Special Food Streets and Towns

227 贰 名特餐饮企业与店铺
Famous Special Food Enterprises and Restaurants

245 第五篇 资阳美食名人与典故
Chapter Five   Celebrities and Allusions of Ziyang Cuisine

247 壹 古代名人与美食典故
Ancient Celebrities and Food Allusions

255 贰 近代至今名人与美食典故
Modern Celebrities and Food Allusions

269 第六篇 资阳名特酒品与茶饮
Chapter Six   Ziyang Famous Special Liquors and Tea Drinks

271 壹 名特酒品
Famous Special Liquors

279 贰 名特茶饮
Famous Special Tea Drinks

Chapter One

THE HISTORY AND DEVELOPMENT OF ZIYANG CUISINE

Ziyang, located in the mid-west of the land of abundance and the west bank of the Tuojiang River, is a city closely connected with Chengdu and Chongqing. Ziyang has a long history and rich cultural connotation. As early as 35,000 years ago, the ancient "Ziyang Man" lived and multiplied here, creating the history of human civilization in Tuojiang River Basin. Later, Ziyang bred Chang Hong, the teacher of Confucius, Wang Bao and Dong Jun of the Han Dynasty, and many other outstanding figures. Thus Ziyang was honored as "the original homeland of Shu people, the hometown of Three Sages". Nowadays, based on the resource endowment, Ziyang has formed a characteristic industrial cluster including food. With its long history, rich cultural deposits and distinctive style, Ziyang cuisine has become the unique name card of Ziyang City in the new era.

第一篇 资阳

美食的历史沿革与发展状况

资阳，地处天府之国的中西部、沱江西岸，是一座与成都、重庆紧密相连的城市。资阳历史悠久，文化内涵丰厚。早在三万五千年前，古老的『资阳人』就在这里繁衍生息，创造了沱江流域的人类文明史。此后，资阳孕育了孔子之师苌弘、汉代王褒、董钧等众多杰出人物，被誉为『蜀人原乡、三贤故里』。如今，资阳立足资源禀赋，形成了包括食品在内的特色产业集群，资阳美食更以其悠久的历史源流、丰富的文化底蕴、鲜明的风格特色成为新时代资阳市活色生香的独特城市名片。

安岳紫竹观音
Zizhu Guanyin of Anyue County

HISTORY OF ZIYANG CUISINE
资阳美食的历史沿革

壹 I

资阳建置始于汉建元六年（公元前135年），初置资中县，隶犍为郡。南北朝北周时改为资阳县，因县城在资水（沱江）之北而得名。隋开皇三年（公元583年），资阳县隶资州。元至正二十二年（公元1362年），废资阳县，地属简州。明成化元年（公元1465年）复设资阳县。清雍正五年（公元1727年），资阳县改属资州。1949年中华人民共和国成立后，资阳县属川南行署资中专区（后更名为内江专区、内江地区）。1985年，资阳县属内江市。1993年，资阳县改为县级资阳市，由内江市代管。1998年，设立资阳地区，辖安岳、乐至两个县和资阳、简阳两个县级市。2000年，撤销资阳地区和县级资阳市，设立地级资阳市和资阳市雁江区。2016年5月，简阳从资阳代管调整为成都市代管。目前，资阳市下辖雁江区、安岳县、乐至县。其中，乐至县在汉代属蜀地郡，隋置普慈县，唐武德三年（公元620年）设立乐至县，因其东有乐志池而得名；安岳县在汉代属犍为郡地，北周建德四年（公元575年）时置普州，历代有废置，清雍正六年（公元1728年）复设安岳县。在漫长的历史发展过程中，资阳作为天府之国中西部的重镇和交通要冲，创造和积累了底蕴深厚、特色突出的美食文化，又结合优质特产食材的生产加工等，形成了独具特色的资阳美食产业，丰富了川菜文化内涵、促进了川菜产业发展。

Ziyang was established in the 6th year of Jianyuan Period of the Han Dynasty (135 B.C.). It was at first set as Zizhong County which was affiliated to Qianwei Prefecture. As the county was located in the north of Zishui River (Tuojiang River), it was renamed as Ziyang County in the Northern Zhou Dynasty of the Southern and Northern Dynasties. In the 3rd year of Kaihuang Period of the Sui Dynasty (583A. D.), Ziyang County was affiliated to Zizhou State. In the 22nd year of Zhizheng Period of the Yuan Dynasty (1362A. D.), Ziyang County was abolished, and the region belonged to Jianzhou State. In the 1st year of Chenghua Period of the Ming Dynasty (1465 A. D.), Ziyang County was rebuilt. In the 5th year of Yongzheng Period of the Qing Dynasty (1727 A. D.), Ziyang County was adjusted to under the jurisdiction of Zizhou State. After the founding of the People's Republic of China in 1949, Ziyang County was affiliated to Zizhong Prefecture of South Sichuan Administrative Office (later renamed as Neijiang Prefecture). In 1985, Ziyang County belonged to Neijiang City. In 1993, Ziyang County was changed to County-level Ziyang City, which was managed by Neijiang City. In 1998, Ziyang Prefecture was established, governing Anyue and Lezhi Counties and Ziyang and Jianyang County-level Cities. In 2000, Ziyang Prefecture and County-level Ziyang City were abolished; Prefecture-level Ziyang City and Yanjiang District of Ziyang City were established. In May, 2016, Jianyang was adjusted to under the jurisdiction of Chengdu City from Ziyang. At present, Ziyang City has jurisdiction over Yanjiang District, Anyue County, and Lezhi County. Among them, Lezhi County belonged to Shu Prefecture in the Han Dynasty and was set as Puci County in the Sui Dynasty. In the 3rd year of Wude Period of the Tang Dynasty (620 A. D.), Lezhi County was rebuilt and named as there was a Lezhi Pond in the east. Anyue County belonged to Qianwei Prefecture in the Han Dynasty. It was set as Puzhou State in the 4th year of Jiande Period of the Northern Zhou Dynasty (575 A. D.). There are abolishments and establishments in different dynasties. In the 6th year of Yongzheng Period of the Qing Dynasty (1728 A. D.), Anyue County was restored. In the long history of development, as the key town and transportation artery of the midwest of the land of abundance, Ziyang has created and accumulated a profound and distinctive food culture. In addition, combined with the production and processing of high-quality specialty food materials, a unique food industry has been formed in Ziyang, which enriches the cultural connotation of Sichuan cuisine and promotes the development of Sichuan cuisine industry.

1 PREHISTORIC TIMES 史前时期

"资阳人"头骨化石、鲤鱼桥遗址、临江窑遗址等，是了解史前时期资阳地区饮食发展的重要窗口。1951年建设成渝铁路时，在资阳火车站以西约1.5公里处发现了人头骨化石1件、骨锥1件和大量共生的化石动物群。这件人头骨化石被正式命名为"资阳人"，已具有晚期智人特征；所发现的骨锥以三棱形骨片磨成，尖端很短，考古学家认为是用刮削方法制成，并推断资阳人是以打制石器作为主要劳动工具。在化石发现地点同一侧右岸上游250~300米的探坑内，还发现了与哺乳动物化石等共生的多件打制石器，考古学家认为是资阳人及其群体所遗。这些出土文物表明，在距今约数万年至10余万年间的资阳大地上，早期"资阳人"群体已经在此繁衍、生息、渔猎，开展了系列生产劳动及饮食活动，创造了资阳的历史与文化。

The skull fossil of "Ziyang Man", Liyuqiao Site, Linjiang Kiln Ruins and so on are important windows to understand the food development of Ziyang area in prehistoric times. When the Chengdu-Chongqing Railway was built in 1951, a human skull fossil, a bone awl and a large number of symbiotic fossil fauna were found about 1.5 kilometers west of Ziyang Railway Station. The human skull fossil, officially named "Ziyang Man", already bears the characteristics of late homo sapiens. The bone awl found is ground by prismatoidal sclerite. It has short tip. The archaeologists believed that it was made by scraping and concluded that the Ziyang Man used chipped stone tools as their main labor instruments. In the test pit 250 ~ 300 meters upstream the right bank on the same side of the fossil site, a number of chipped stone tools symbiosed with mammalian fossils were also found, which archaeologists believed were left by the Ziyang Man and his group. These indicate that early "Ziyang Man" group multiplied, lived, fished and hunted, carried out a series of productive labor and dietary activities, and created the history and culture of Ziyang on the land of Ziyang of about dozens of thousand years ago to more than 100 thousand years ago.

1973年和1980年，考古专家相继在雁江区鲤鱼桥河床出土了多种哺乳类动物化石、打制石器和大量的乌木、树叶及种子标本。其中，打制石器主要包括石片、砍伐器、刮削器等，是数万年前当地先民

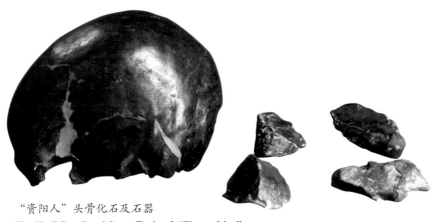

"资阳人"头骨化石及石器
The Skull Fossil and Stone Tools of "Ziyang Man"

的生产工具,表明当时的食材来源是狩猎和采集并重。此外,在鲤鱼桥遗址中还发现了大量陶片,包括碗、罐、尖底器等,考古专家认为它们是新石器时代的文化遗物。

In 1973 and 1980, archaeological experts successively unearthed a variety of mammal fossils, chipped stone tools and a large number of ebony, leaf and seed specimens in the riverbed of Liyu Bridge in Yanjiang District. Among them, chipped stone tools mainly include flakes, chopping tools and scraping tools, which are the production tools of local ancestors of tens of thousands of years ago, indicating that the food at that time was from both hunting and gathering. In addition, a large number of pottery fragments, including bowls, pots, bottom-pointed vessels, etc. have been found at Liyuqiao Sit, which archaeologists believe are cultural relics of the Neolithic Age.

进入新石器时代后,资阳先民们的生产力水平、活动范围有了重大发展,逐渐形成原始农业经济。资阳出土的大量新石器时代文物,如临江窑旧址出土的陶器、陶罐和石斧,反映了当时人们在陶器制作、农业生产、饮食生活等方面的进步。

After entering the Neolithic Age, the level of productivity and range of activities of the ancestors of Ziyang had a significant development, and gradually formed a primitive agricultural economy. A large number of Neolithic relics unearthed in Ziyang, such as earthenware, pottery pots and stone axes unearthed at Linjiang Kiln Ruins, reflect the then people's progress in pottery making, agricultural production, diet and life, etc.

商周时期 SHANG AND ZHOU DYNASTIES 2

今雁江区、乐至县在商周时期属于古蜀文化的范畴,而安岳则属巴蜀分治之地,在这一时期都深受巴蜀文明的影响。据历史学家研究,早在四五千年前,包括资阳在内的成都平原及周边丘陵、山地的农业已得到初步开发。到西周时代,蜀地已成为当时全中国农业先进的富庶之区。春秋战国时期,蜀国大规模兴修水利,促进了农业的长足进步,而资阳的农业发展与沱江密不可分。据《尚书·禹贡》载:"岷山导江,东别为沱。"这是关于沱江贯穿资阳的早期文献资料。由于有沱江水的滋养,资阳农业得到良性发展,也为民众饮食提供了物质基础。

Today's Yanjiang District and Lezhi County belonged to the category of Ancient Shu Culture in Shang and Zhou Dynasties, while Anyue County belonged to two cities of Ba and Shu. During this period, they were deeply influenced by Bashu Civilization. According to the research of historians, as early as four or five thousand years ago, agriculture in Chengdu Plain, including Ziyang, and the surrounding hilly areas had been preliminarily developed. By the time of the Western Zhou Dynasty, Shu had become a prosperous area with advanced agriculture in China. During the Spring and Autumn Period and the Warring States Period, the Kingdom of Shu built water conservancy projects on a large scale, which promoted the great progress of agriculture. Ziyang's agricultural development and Tuojiang River are inseparable. It is recorded in *Shangshu · Yugong* that "(Dayu) began to dredge water from Minshan Mountain, and formed Tuojiang River in the east". This is the early document materials about Tuojiang River running through Ziyang. Thanks to the nourishment of Tuojiang River, Ziyang agriculture has been well developed, and it also provides the material basis for people's diet.

3 HAN AND JIN DYNASTIES 汉晋时期

秦灭巴蜀以后，逐渐向巴蜀地区大量移民，带来了中原的农业技术和饮食习俗，促进了包括资阳在内的巴蜀地区农业生产、饮食生活的发展。这一时期，资阳美食的发展主要体现在4个方面：

After Qin destroyed Bashu, a large number of Qin people gradually migrated to Bashu area, bringing the agricultural technology and dietary customs of Central Plains, and promoting the development of agricultural production and diet and life in Bashu area including Ziyang. During this period, the development of Ziyang cuisine is mainly reflected in the following four aspects.

陶哺乳俑（东汉）
资阳市南市镇崖墓出土
Potter Breast-Feeding Figurine (Eastern Han Dynasty) Unearthed From a Cliff Tomb in Nanshi Town, Ziyang City

农耕经济的发展提供了种类丰富的植物类食材。其中，粮豆类有麦、黍、芋、大豆等，果蔬及调料有桃、李、梨、柑橘以及莲藕、姜、蒜等。汉代时，由于铁制农具与牛耕的出现并普及，很多地区开始精耕细作，农作物种类不断丰富。汉代资阳人王褒撰写的《僮约》有许多关于农业耕种和收获食材的记载，如文中规定，新买奴僮必须"种姜养羊""种瓜作瓠""园中拔蒜""池中掘荷""收芋"，"十月收豆、麦、窖芋""植种桃、李、梨、柿、柘、桑"，呈现出一幅比较发达的农业生产图景。在2005年和2006年发掘的雁江区兰家坡汉墓中就出土了多件劳作陶俑和陂塘陶制模型，2010年雁江区狮子山崖墓中出土了陶井1件，也在一定程度上反映了当时精耕细作的状况。至魏晋时期，种植历史久远的蜀芋已经成为一种重要的粮食作物。北魏贾思勰著《齐民要术》载："蜀汉既繁芋，民以为资。"

Firstly, the development of farming economy provided a rich variety of plant foods. Among them, grains and legumes include wheat, millet, taro, soybean and so on; fruits and vegetables and condiments include peach, plum, pear, citrus and lotus root, ginger, garlic and so on. During the Han Dynasty, due to the emergence and popularization of iron farming tools and ox-ploughing, intensive farming began in many areas and crop varieties were constantly enriched. In *Tongyue* written by Wang Bao of the Han Dynasty, there are many records of farming and harvesting of food materials. For example, it is stipulated that the newly bought slave servant must "plant ginger and breed sheep", "grow melon and make gourd", "pull out garlic in the garden", "dig lotus root in the pool", "harvest taro", "reap beans and wheat, and pit taro in October", "plant peach, plum, pear, persimmon, cudrania and mulberry", presenting a relatively developed agricultural production picture. In the Han Tomb of Lanjiapo in Yanjiang District excavated in 2005 and 2006, a number of labor pottery figures and Beitang pottery models were unearthed. In 2010, a pottery well was unearthed in the Cliff Tomb of Shizishan in Yanjiang District, which to some extent reflects the intensive farming conditions at that time. By the Wei and Jin Dynasties, Shu taro, which had a long planting history, had become an important food crop. It is recorded in *Qi Min Yao Shu* written by Jia Sixie of the Northern Wei Dynasty that "taro was widely planted in the kingdom of Shu during the Han Dynasty, and people regarded it as an important resource".

渔猎畜牧等辅助经济的发展，为人们提供了水产及禽畜等肉类食材。当时的水产食材，除了在江河溪流中捕捞外，还来自塘堰的家养鱼。《僮约》言"结网捕鱼""入水捕鱼""垂钓"和"调治马户"。"户"在此指"水门"。当时蜀地通常利用沟、溪，设置水闸来养鱼，以供食用。这一时期，狩猎依然是资阳先民们的主要经济生活之一。《僮约》言"徼雁弹凫""登山射鹿"等。此外，在雁江区兰家坡汉墓等墓葬中都发现了多件陶猪、陶鸡、陶狗、陶马等器物，表明此时饲养的牲畜也成为人们重要的肉食来源。

Secondly, the development of auxiliary economy, such as fishing, hunting, and livestock rearing, provided people with aquatic products, livestock and other meat materials. At that time, aquatic food materials, in addition to fishing in rivers and streams, also came from breeding fish in pond. *Tongyue* includes the content of "fishing with net", "fishing in water", "going fishing" and "adjusting water gate". At that time, ditches and streams were usually used to raise fish for food in Shu area. In this period, hunting was still one of the major economic activities of Ziyang ancestors. *Tongyue* describes the pictures of "hunting wild goose and catching wild duck", and "climbing mountain and shooting deer", etc. In addition, many pottery pigs, chickens, dogs, horses and other artifacts were found in the Han Tomb of Lanjiapo in Yanjiang District and in other tombs, indicating that livestock raised at this time also became an important source of meat for people.

茶酒的饮用。王褒在其《僮约》中写到，要求奴僮"烹茶尽具""武阳买茶"，这是世界上最早关于买茶和饮茶的文献，表明当时资阳已有饮茶习惯和专门的茶具。资阳酿酒、饮酒历史悠久，据《僮约》中言，严格限制奴僮饮酒、"不得倾杯覆斗"，说明当时包括奴仆在内的人群都可以饮酒，酒已成为社会各阶层的消费品。

Thirdly, the drinking of tea and liquor. Wang Bao wrote in his *Tongyue* that slave servant was required to "boil tea and clean tea set" and "buy tea in Wuyang". It is the earliest document on buying and drinking tea in the world, indicating that the habit of drinking tea and special tea sets existed in Ziyang at that time. Ziyang has a long history of liquor brewing and drinking. In *Tongyue*, drinking by slave servant is strictly restricted and "(the servant) cannot let the bottom of the cup be seen", indicating that people including servants could drink at that time and liquor had become the consumer goods of all levels of society.

饮食器具种类较为丰富。在雁江区兰家坡汉墓、狮子山崖墓中，均发现了众多陶制、铜制、铁质类饮食器具，主要有陶钵、陶罐、陶耳杯、陶甑、陶灶、陶盆、陶碗、陶案，以及铜耳杯、铜釜、铜筷、铁釜等。其中，在陶案上和铜釜中都发现了动物骨骼。

Finally, there were relatively rich varieties of eating utensils. In the Han Tomb of Lanjiapo and the Cliff Tomb of Shizishan in Yanjiang District, many pottery, copper and iron eating utensils were found, mainly including pottery bowl, pottery pot, pottery ear cup, pottery rice steamer, pottery stove, pottery basin, pottery table and copper ear cup, copper kettle, copper chopsticks, iron kettle, etc. Among them, animal bones have been found on the pottery table and in the copper kettle.

4 TANG AND SONG DYNASTIES | 唐宋时期

唐代时，包括资阳在内的四川盆地中部地区农业发展较快。根据郭声波《四川历史农业地理》一书统计，唐代天宝年间，资阳人口约10.477 5万，耕地约1.126 3万顷，处于整个四川的中上水平。这一时期，粮食类食材仍然以麦、黍、芋等为主。其中，芋的种植地区已遍布整个四川盆地及盆周山区。在果蔬及其他植物性食材方面，柑橘和甘蔗的种植尤为普遍。柑橘在资阳的种植历史悠久，至唐代已遍及四川盆地中东部的低山丘陵地区，其中的主要产地有资州、普州和简州等。据《资阳花溪志》载，公元806年，资阳的贡品中就有柑子。宋代时，四川甘蔗的种植遍布涪江、沱江流域的资州、遂州、梓州、汉州等地，据《太平寰宇记·资州》（卷76）所载，物产有高良姜、甘蔗。当时的这一区域制糖业已相当发达，是全国重要的产糖基地。此外，当地还有很多其他特色农产品，如宋代王象之《舆地纪胜·普州》（卷158）所载："郡土硗瘠，无珍异之物，惟铁山枣、崇龛梨、天池藕，三物皆陈希夷所种。"可见，天池藕在宋代时已很有名。

During the Tang Dynasty, agriculture developed rapidly in central Sichuan Basin including Ziyang. According to Guo Shengbo's *Historical Agricultural Geography of Sichuan*, during the Tianbao Period of the Tang Dynasty, Ziyang had a population of about 104,775 and cultivated land of about 112.63 million qing, which was in the middle and upper level of Sichuan. During this period, the main grain food materials were wheat, millet and taro. Among them, the planting area of taro had spread throughout the whole Sichuan Basin and the mountains around. As for the plant food materials such as fruits and vegetables, citrus and sugar cane were especially popular. Citrus has a long history of planting in Ziyang. By the Tang Dynasty, it had spread throughout the low hills and hilly areas in the middle and east of Sichuan Basin, among which the main producing areas are Zizhou, Puzhou and Jianzhou. According to the records *Ziyang Huaxi Annals*, citrus was one of the tributes of Ziyang in 806. During the Song Dynasty, sugar cane cultivation in Sichuan had been spread all over Zizhou, Suizhou, Zizhou, Hanzhou and other places in the Tujiang and Tuojiang river basins. The recorded products, the 76th Volume of *Universal Geography of the Taiping Era · Zizhou*, include Gaoliang ginger and sugar cane. At that time, the sugar industry in this region was quite developed and was an

important sugar production base in the country. In addition, there were many other local agricultural products. For example, it is recorded, the 158th Volume of *The Record of Scenic Spots Across the Country · Puzhou* by Wang Xiang of the Song Dynasty that "the land here is poor, with no precious things; only Tieshan jujube, Chongkan pear and Tianchi lotus root are all planted by Chen Xiyi". It can be seen that Tianchi lotus root was already very famous in the Song Dynasty.

唐宋时期，资阳的酿酒业、制盐业也有了较快发展。据学者研究，唐中叶至宋末时，资阳酿酒业以专卖为主，征税为辅。1077年以前，在今资阳境内分设5处酒务实行专卖，并允许三江、赖琬、丹山等镇民众酿酒以收税，全县酒税岁额两万贯。1130年，川内允许百姓经营酒业，官民分设酿酒作坊和酒店。四川制置使胡世将在资州、普州等地创设清酒务。在制盐业方面，据《四川通史》统计，唐开元二十五年（公元737年），四川有10州产盐，盐井90所，集中于四川盆地的中部丘陵地区，其中资州有28所，梓州、普州等5州共有38所。

During the Tang and Song Dynasties, Ziyang's liquor brewing industry and salt making industry also developed rapidly. According to scholars 'research, from the middle period of Tang to the end of Song, Ziyang's liquor brewing industry was dominated by monopoly and supplemented by taxation. Before 1077, in present-day Ziyang, five departments were set up to sell liquor exclusively, and the people in Sanjiang, Laiwan, Danshan and other towns were allowed to brew liquor for the purpose of collecting taxes. The annual liquor tax of the county was 20, 000 guan. In 1130, people are allowed to do liquor business in Sichuan, and the officials and citizens set up liquor workshops and hotels separately. Sichuan Military Commissioner Hu Shijiang set up pure liquor departments in Zizhou, Puzhou, etc. In the salt industry, according to the statistics of *Sichuan General History*, in the 25th year of Kaiyuan Period of the Tang Dynasty (737 A. D.), there were 10 states producing salt in Sichuan, 90 salt wells concentrated in the hilly areas of the central Sichuan basin, among which there were 28 wells in Zizhou and 38 wells in 5 states including Zizhou and Puzhou.

唐宋时期，资阳地区宗教类石刻造像迎来发展高峰，其中以安岳石刻、乐至报国寺、资阳半月山大佛等最具代表性。安岳石刻始于南北朝，盛于唐宋，有石刻230处10万余尊，高3米、雕技精湛的紫竹观音更

乐至报国寺树抱佛　　Tree-Hugging Buddha in Baoguo Temple of Lezhi County

是被誉为"东方维纳斯"。乐至报国寺始建于隋开皇二年（公元582年），拥有国内寺院少见的众多千面玉佛像。这一时期大量的宗教石刻、碑记、佛像壁画表明，当时的资阳已是西南地区重要的宗教文化传播中心，在经济、社会和民众生活等方面已有较高的发展水平，这些也对资阳美食的发展产生了一定促进作用。

During the Tang and Song Dynasties, the religious stone carvings in Ziyang area ushered in the peak of development, among which Anyue Sculpture Carvings, Lezhi Baoguo Temple, Ziyang Half Moon Mountain Buddha were the most representative. Anyue Sculpture Carvings began in the Northern and Southern Dynasties and were popular in the Tang and Song Dynasties. There are more than 100,000 stone carvings in 230 places. The exquisite Zizhu Guanyin is known as "Venus in the East". Built in the 2nd year of Kaihuang Period of the Sui Dynasty (582 A. D.), Lezhi Baoguo Temple has many Jade Buddha of thousand faces rarely seen in domestic temples. During this period, there were a large number of religious stone carvings, inscriptions on stone tablets and frescoes of Buddha statues showing that the then Ziyang already became an important religious cultural communication center in southwest China, and had a high level of development in economy, society and people's life, which also promoted the development of Ziyang cuisine.

5 MING AND QING DYNASTIES 明清时期

这一时期，大量移民进入资阳，带来了新食材，饮食品生产、加工不断增加，资阳美食有了新的发展并逐步走向高峰，为后世餐饮业的发展奠定了良好基础。

During this period, a large number of immigrants entered Ziyang, bringing new food materials. The production and processing of food continued to increase, and Ziyang cuisine had a new development and gradually reached the peak, which has laid a good foundation for the development of catering industry of later generations.

元末明初和明末清初两个阶段，资阳与四川的其他地区一样受到战乱摧残，人口锐减、土地荒芜，而随着政局的稳定，四川迎来了大规模"湖广填四川"的移民浪潮。其中，大量来自湖北、湖南等地的移民在资阳安居。据《资阳花溪志》载，明永乐七年（公元1409年），湖北麻城孝感乡卓用祥迁居资阳清水河，后住杨花场，人口众多，号称卓半乡。这些移民为资阳带来了大量劳动力和先进的农业生产技术，促进了农业发展和食材的丰富。据乾隆时期的《安岳县志》载，当时的安岳境内，"开垦殆遍，几于野无旷土矣。农民各勤稼穑，务蚕桑，……四方侨寓，复多秦、粤、吴、楚之人，始则佃地而耕，终则携家落业，虽曰客民，同于土著矣"。清代时，资阳民众的食材种类也很丰富。据嘉庆时期的《资阳县志》载，当地日食三餐皆稻米，遇荒年及贫者间用面、菽、粟、黍、芋、苕等。另据其所列物产统计，谷类食材有12种，粟类食材有高粱、黍、粟等6种，菽类食材有黄豆、黑豆等11种，蔬菜有葱、蒜、韭等19种，瓜果类有南瓜、冬瓜、桃等34种。清代后期，资阳众多的优质农产品还大量销往成都，据清末傅崇矩《成都通览》载，1909年"成都之外来农业陈列品"中，资阳主要有晚花蜀麦、大胡豆、朱砂豌豆、红花等，乐至主要有红枣、红花、藕粉，安岳主要有糯稻、白果等。此时，乐至的天池藕粉被称为"蜀中第一品"。此外，原产于美洲的玉米、番薯等食材也随着移民入川而传入，在资阳的中下层民众饮食中开始占有一席之地。据道光时期的《乐至县志》载："有玉麦，即玉蜀也，俗称包谷。……宜作酒，贫户磨面煎饼，充足日食。""番薯，土人谓之红苕，可耕可饭。县自中人产，无不栽植，冬藏于土窖，足供数月之食。"番薯，即红薯、红苕，在当地也得到广泛种植和食用。玉米、番薯成为当时荒年缓解粮食不足的重要食材。

During the two periods of late Yuan and early Ming Dynasties, and late Ming and early Qing Dynasties, Ziyang, like other regions in Sichuan, was ravaged by wars, with a sharp decrease in population and a barren land. With the stabilization of political situation, Sichuan received a wave of immigrants called "Hu Guang Filing Sichuan". A large number of immigrants from Hubei, Hunan and other places settled and lived in Ziyang. According to the records of *Ziyang Huaxi Annals*, in the 7th year of Yongle Period of the Ming Dynasty (1409 A. D.), Zhuo Yongxiang from Xiaogan township of Macheng City, Hubei Province, moved to Qingshuihe River of Ziyang, and later lived in Yanghua Field, which also known as Half Zhuo Township because the large population of Zhuo's. These immigrants brought a large number of labor force and advanced agricultural technology to Ziyang, which promoted the development of agriculture and the abundance of food materials. According to *Anyue Annals* of Qianlong Period, in Anyue at that time, "the land was exploited completely with few wastelands in the wild. Farmers were diligent in farming and accumulation and breeding silkworm and growing mulberry… People from different regions, mostly from Qin, Yue, Wu, Chu, came to live here. At first they rented lands to farm and settled down with their families in the end. Although called outsiders, they lived as natives here." In the Qing Dynasty, the people of Ziyang had a rich variety of food materials. According to *Ziyang Annals* of Jiaqing Period, people there ate rice in three meals; in lean years and poor families, people sometimes ate flour, beans, millet, panicum, taro, potato and so on. In addition, according to the statistics of listed products, there are 12 kinds of cereals, 6 kinds of millet (sorghum, panicum, millet, etc.), 11 kinds of beans (soybean, black bean, etc.), 19 kinds of vegetables (shallot, garlic and leek, etc.), and 34 kinds of melons and fruits (pumpkin, wax gourd, peach, etc.). In the late Qing Dynasty, numerous high-quality agricultural products of Ziyang were sold to Chengdu in large quantities. In Fu Chongju's work *Chengdu Overview*, among the "Agricultural Exhibits Not from Chengdu" in 1909, there were late-flowering Shu wheat, big broad bean, cinnabar pea, safflower, etc. in Ziyang, red dates, safflower, lotus root porridge, etc. in Lezhi, and glutinous rice, ginkgo in Anyue. At that time, Tianchi Lotus Root Porridge of Lezhi was known as "The Best of Shu". Furthermore, corn, sweet potato and other food materials originated in America was introduced into Sichuan by immigrants, and began to take a place in the diet of the middle and lower classes in Ziyang According to the records of *Lezhi Annals* of Daoguang Period, "corn…is suitable for liquor making. The poor bake corn flour pancake to supplement the diet." "Sweet potato can be planted and eaten. Almost all the people of the county plant it. They can be stored in crypt in winter for several months' eating." Sweet potato was also widely cultivated and eaten in local areas. At that time, corn and sweet potato became important food materials to alleviate food shortage in lean years.

此时，资阳地区的糖、酒等生产与销售持续发展，豆瓣等调味品生产逐渐兴起。清代时，四川种植甘蔗的范围遍及内江、资中、资阳、安岳等21个厅、州、县，尤以资中、资阳等地发展最快，产量最多。据《资阳县志》载，清咸丰十年（公元1860年），资阳县境沿沱江两岸广种甘蔗，年产红糖3万桶，约合1 800万斤。清代时，资阳酒的生产与销售也是更上一层楼。据嘉庆时期的《资阳县志》载："酒酿于家者，有窖酒、常酒；沽于市者有烧酒大曲、惠泉老酒，酒市所在皆有。"可见，民间酿酒之风盛

行，而以伍市干酒最为有名。《成都通览》中记载了资阳出产的烧酒、高粱、蔗糖等在成都十分畅销。此外，清乾隆三年（公元1738年），临江寺豆瓣创始人聂守荣在临江寺场开设义兴荣酱园。此后不断发展，整个临江寺场逐渐成为制作、贩运豆瓣的一条街，临江寺豆瓣销往成都及其他地区。

打造资阳老酒　Making Ziyang Aged Liquor

During this period, the production and sales of sugar and liquor in Ziyang area continued to develop, and the production of chili bean paste and other condiments gradually rose. In the Qing Dynasty, sugarcane was planted in 21 different regions in Sichuan, including Neijiang, Zizhong, Ziyang and Anyue, among which Zizhong and Ziyang had the fastest development and the largest output. According to the records of *Ziyang Annals*, in the 10th year of Xianfeng Period (1860 A.D.), sugarcane was widely planted along the Tuojiang River in Ziyang County, with an annual output of 30,000 barrels of brown sugar, about 18 million jin in total. In the Qing Dynasty, the production and sales of Ziyang liquor also reached a higher level. According to *Ziyang Annals* of Jiaqing Period, "The liquors brewed at home include and daily liquor; those sold and popular at the market include daqu, Huiquan aged liquor." So the folk brewing at that time was prevailing. And the most famous liquor was Wushi Dry Liquor. It is recorded in *Chengdu Overview* that the liquor, sorghum and sucrose produced in Ziyang were very popular in Chengdu. In addition, in the 3rd year of Qianlong Period of the Qing Dynasty (1738 A.D.), Nie Shourong, the founder of Linjiangsi chili bean paste, set up Yixinrong Sauce and Pickle Factory in Linjiangsi Temple Field. Since then, the whole Linjiangsi Temple region gradually became a street for producing and selling chili bean paste and Linjiangsi chili bean paste was sold to Chengdu and other areas.

这一时期，资阳城乡场镇商业逐渐发达，推动了餐饮业发展。乡间的场镇历来都是集市贸易的场所，也成为民间饮食业活动的中心。此时，资阳地区形成了众多场镇，如乾隆时期的《安岳县志》中就列举了永兴场、龙台场、毛家场等84处，记载圆坝场连接东西、"商贾便之"，千佛场"商民繁庶"。可见，当时场镇是区域性商贸重要集散地，汇聚了大量人流和物流，为餐饮业的发展提供了巨大空间。在城市里，饮食生活则更丰富多彩。据道光时期的《乐至县志·风俗》载："城市则有茶酒店、面店、肉铺、生药、果子……油酱、食米、见成饮食等铺坐列贩卖，以便日用。"各类餐饮店众多，民众饮食消费兴旺。《资中竹枝词》中写到雁江两岸"酒楼灯火耀通衢，城郭参差入书画"，可以想见当时资阳城区餐饮业的兴盛。

Moreover, the commerce of urban and rural areas in Ziyang gradually developed at this period, promoting the blossom of catering industry. The towns in countryside have always been the place of market trade, and also the center of folk catering activities. At this time, Ziyang formed a number of towns, such as Yongxing Field, Longtai Field, Mao's Field and other 84 towns listed in *Anyue Annals* of Qianlong Period, among which Yuanba Field connected east and west to "make the business convenient", and Qianfo Field was "prosperous in trading". It can be seen that the town was an important distribution center for regional trade at that time, gathering a large number of people and logistics and providing a huge space for the development of catering industry. In the city, the diet life was more colorful. According to the records in *Lezhi Annals · Customs* of Daoguang Period, "there are tea house and restaurant, noodle shop, butcher shop, crude drug, fruit…oil, sauce, rice and ready-made meal stores convenient for daily life in the city." Various kinds of restaurants stood in great numbers, and people's food consumption was flourishing. It was written in *Zizhong Zhuzhi Ci* that, on both sides of Yanjiang River, "The lights in the restaurants lightened the thoroughfares, and the cities were irregularly scattered like a picture." It can be seen that the catering industry in Ziyang city was prosperous at that time.

民国时期 | REPUBLIC OF CHINA PERIOD — 6

进入民国时期，资阳地区主要种植粮食、甘蔗和蔬果，民众主食也主要是水稻、玉米、红薯、小麦、豌豆等，养殖猪、羊、牛、兔和小家禽等作为肉食原料，蔬菜品种丰富。与此同时，该地区有意识地引进优良农作物品种，各种饮食类手工业也不断发展，不仅丰富了食材种类、特色愈加鲜明，也促进了饮食店铺的增多和相关行帮组织的产生。

In the period of the Republic of China, grain, sugar cane, vegetables and fruits were mainly cultivated in Ziyang area. The staple foods of the people were rice, corn, sweet potato, wheat and pea. Pigs, sheep, cattle, rabbit and small poultry were cultivated as raw materials for meat. And there were rich varieties of vegetables. At the same time, Ziyang consciously introduced superior crop varieties, and a variety of catering handicraft industries were unceasingly developed, which not only enriched the variety and distinctive characteristics of food materials, but also promoted the increase of food stores and the emergence of related trade organizations.

这一时期，资阳地区引进异域农作物品种取得了一定成效。如安岳县龙台场人邹海帆于1929年将柠檬引入家乡，其父邹江亭精心培育出安岳柠檬，使柠檬在安岳落户并广泛种植。据《资阳县志》载，1940年资阳县引进POT2878爪哇蔗11 900斤，种植8.1亩（1亩≈0.0667公顷，下同），产蔗63 350斤，亩产7 833斤；1946年，资阳县引进晚熟优质高产南瑞苔进行试种，亩产达4 000斤以上，此后，该品种延续种植30年。除了引种，当地还大量加工特色食材，如粉条、挂面等。据民国时期的《乐至县志又续·物产志》载，"豆粉、条粉"的制法，即以豌豆磨浆，去渣、取浆放入木桶沉淀，"取粉出晒干，为欠粉，再用人工做成面丝，为条粉。每升豌豆得粉二十七八两""绿豆、红苕，均可取粉"。该书还记载了挂面制法：小麦磨粉、过筛，"入盆用水与菜油及盐做成团，干湿合宜取出，置案盘成条，上竹签置槽内。次日上木架，挂为细丝，风干、收成把。"由于手工制作挂面较为复杂，当时已采用机器加工："近有机器作者，成面略省劳，……通县相习，几成大宗。"此外，资阳的特产食材还有许多，如民国时期的《资阳县志稿·物产》载，有胡豆、豌豆、仙米、白米、豆瓣、梨、橘橙等。

During this period, the introduction of exotic crop varieties in Ziyang region achieved certain results. For example, Zou Haifan, a native of Longtai Field of Anyue County, introduced lemon to his hometown in 1929. His father Zou Jiangting carefully cultivated Anyue Lemon, so that the lemon settled in Anyue and was widely planted. According to *Ziyang Annals*, in 1940, Ziyang County introduced POT2878 Java sugar cane 11,900 jin, planted 8.1 mu (1mu≈0.0667 hectare, the same below), produced 63,350 jin of sugarcane, per mu yield 7,833 jin. In 1946, Ziyang County introduced late-maturing good quality and high yield Nanrui sweet potato for trial planting. The yield per mu reached more than 4,000 jin. Since then, the variety continued to be planted for 30 years. In addition to introducing new varieties, local people also processed a lot of special food materials, such as vermicelli and noodles. In *Continuation to Lezhi Annals · Products Records* of the Republic of China era, it wrote the making procedures of "bean flour and vermicelli". That is, grinding the peas into thick liquid, removing slag, and putting the thick liquid into the wooden barrel to precipitate, "taking out and drying the powder in the sun to get starch, and then making it into silk noodles called vermicelli. 27 or 28 liang of vermicelli would be got out of per liter of peas." Both mung bean and sweet potato can make vermicelli". This book also recorded the method of making noodles. Namely, grind wheat into flour and sift, "put it into the basin, add water and rape oil and salt to make dough, take out with appropriate degree of dryness and wetness, place it on the copping board and coil into strip, and place bamboo stick in the slot. Put it on the timber frame the next day to hang for strings, air dry, and collect them in bundles". Due to the complexity of hand-made noodles, machine processing was adopted at that time." Recently somebody made noodles with machine and saved labor…Then the whole county learned and nearly became the general trend." Furthermore, there were many special products in Ziyang. For example, *Ziyang Annals · Products* of the Republic of China era recorded broad bean, pea, rice, chili bean paste, pear, orange and so on.

此时，资阳地区的各种饮食类手工业也进入较快的发展阶段。据《资阳县志》等史料记载和统计，1937年前后，资阳县酿酒业共420家，年产酒420万斤，其中，以伍隍乡最多，有40余家，年产约150余万斤。资阳县酱园业著名的有13家，包括城关镇邓伯余、张述明等，临江寺国泰兴、国泰明等；年产豆瓣及各种酱制品120余吨，主要产品有金钩豆瓣、火肘豆瓣、香油豆瓣、麻油豆瓣等10余种，豆瓣酱、芝麻酱、甜面酱、豆腐乳、酱油、醋等，多销于邻县及成渝等地。当时资阳县酿酒业和酱制品业声誉极高，所产伍市干酒、临江寺豆瓣远近闻名。此外，糖果业也较为兴盛。1929年，资阳县城内有3家京果业较为著名，生产樱桃糖、花生糖等17个产品，还曾参加成都劝业展销会。到1937年，全县有糖坊271家，产红糖4.8万余桶（约1.2万余吨）；京果业增至23家（县城11家）。在农村，农民也开始兼营别业，榨油、酿酒、酿醋、酱菜、磨面、碾米、生豆芽、制豆腐等作坊遍及农村各个角落。而据《乐至县志》载，1931年，乐至县谭福庆等人恢复了"盐肘子"即火腿的生产，年产销约2万余只。1932年，乐至商人秦耀光经营藕粉业务，虽在城郊租田500亩，年产鲜藕30余万斤，但天池藕粉仍然供不应求，县城另有10余家生产藕粉。民国时期，乐至县内酱料业生产以酱油、醋、豆豉为大宗，年产量约10万斤，县内著名商户有近10家，永兴乡金玉号的豆豉、高寺乡舒建章酱园的芝麻酱及豆腐乳为著名产品，行销县内及成都等地。

At this time, all kinds of food handicraft industries in Ziyang area also entered a rapid development stage. According to the historical records and statistics of *Ziyang Annals* and other materials, around 1937, there were 420 liquor businesses in Ziyang County with an annual output of 4.2 million jin, among which Wuhuang Township had the most, with more than 40 businesses and an annual output of over 1.5 million jin. There were 13 famous sauce and pickle factories in Ziyang County, including Deng Boyu and Zhang Shuming of Chengguan Town, and Guotaixing and Guotaiming of Linjiangsi Temple. The annual output of chili bean paste and various sauce products was more than 120 tons. The major products included dried shrimp chili bean, ham chili bean, sesame oil chili bean, etc, and over 10 kinds of chili bean paste and sesame paste, sweet soybean paste, preserved bean curd, soy sauce and vinegar, and were mainly sold in neighboring counties and Chengdu, Chongqing, and other places. At that time, Ziyang liquor brewing industry and sauce making industry were of high reputation. Wushi Dry Liquor and Linjiangsi chili bean paste produced here were known far and wide. Besides, confectionery industry was also quite booming. In 1929, there were 3 famous Jingguo businesses which had ever participated in Chengdu Quanye Exhibition in Ziyang, producing 17 products such as cherry candy, peanut candy, etc. By 1937, there were 271 sugar mills in the county, producing more than 48,000 barrels of brown sugar (about 12,000 tons). The amount of Jingguo businesses increased to 23 (11in the county). In rural areas, peasant began to start a side business, oil manufacture, liquor brewing, vinegar making, pickles, flour grinding, rice milling, bean sprouts producing, tofu making and other workshops were throughout the countryside. According to the records of *Lezhi Annals*, in 1931, Tan Fuqing and others in Lezhi resumed production of ham, with an annual output and sales over 20,000. In 1932, Lezhi merchant Qin Yaoguang was engaged in lotus root porridge business. Although 500 mu of fields were tented in the suburban, with an annual output of more than 300,000 jin of fresh lotus roots, yet Tianchi Lotus Root Porridge was still in short supply. And there were over 10 other factories producing lotus root porridge in the county. During the period of the Republic of China, the sauce industries of Lezhi mainly produced soy sauce, vinegar and fermented soybeans, with an annual output of about 100,000 jin. There were nearly 10 famous shops in the county. The fermented soybeans of Jinyu in Yongxing Township, sesame paste and preserved bean curd of Shu Jianzhang Sauce and Pickle Factory in Gaosi Township were famous products and were sold throughout the county and in Chengdu and other places.

此外，这一时期，资阳地区饮食店、茶馆等也有一定发展，满足着城镇居民的消费需求。据《资阳县志》载，1939年~1940年间，仅资阳县城就有饮食店44户、茶馆35户。据《安岳县志》等记载和统计，1933年，安岳县有茶馆241家。其中，县城28家，以一洞天、小潭天等13家茶馆的规模较大、生意十分兴隆；乡镇213家，以龙台、通贤、李家、周礼等7个大场镇为最多。到1948年，全县共有饮食业782户，其中，酒馆、饭馆294户，面食店146户，茶馆342户。

Moreover, during this period, restaurants and tea houses in Ziyang also developed to meet the consumption needs of urban residents. Records of *Ziyang Annals* showed that, from 1939 to 1940, there were 44 restaurants and 35 tea houses in Ziyang County alone. According to records and statistics of Records of *Anyue Annals*, there were 241 tea houses in Anyue County in 1933. Among the 28 tea houses in the county, 13 teahouses were of large scale and ran well, such as Yidongtian, Xiaotantian, etc. There were 213 tea houses in the township and 7 large towns such as Longtai, Tongxian, Li's, Zhouli, etc. had the largest number. By 1948, there were 782 catering businesses in the county, including 294 taverns and restaurants, 146 noodle stores and 342 tea houses.

随着饮食类手工业和饮食店铺的快速发展，资阳产生了数量较多的行帮组织，包括厨工帮、酿酒帮、屠帮、米帮、盐帮、茶馆帮、糖果帮、酱醋帮等，以此进行一些行业自律和行业管理。

With the rapid development of catering handicraft industry and food stores, a large number of trade association organizations emerged in Ziyang, including kitchen staff association, liquor-brewing association, butcher association, rice association, salt association, tea house association, candy association, sauce and vinegar association, etc., to carry out some industry self-discipline and industry management.

CURRENT SITUATION OF THE DEVELOPMENT OF CONTEMPORARY ZIYANG CUISINE
当代资阳美食发展现状

中华人民共和国成立以后至改革开放初期，资阳美食发展尽管经历了一些曲折，但依然取得了较大进步。在农业方面，大力发展农业、促进生产，粮、油、蔗等作物获得了一定增长。如在经济作物种植上，1949年时种植面积约14.93万亩，到1985年扩大至53.74万亩，还建立了资阳玉米原种场等多个国营农场，为资阳美食提供了丰富多样的食材。在食品饮料制造业方面，资阳也有了较大发展。如1957年建立资阳糖厂，1958年建立南湍河糖厂，至1985年，根据工业总产值行业构成统计，食品制造业产值占比21.38%，排名第一，企业83个，产值8 049万元；饮料制造业32家，产值970万元，占比2.58%。这些企业的建立与发展，推动了资阳美食从个体手工业向工业化生产的转变。在餐饮业方面，资阳地区也有一定的发展和转变。据《资阳县志》《安岳县志》等记载，1955年，安岳县有私营饮食业1 301户，比1948年增加66.4%。1956年，全行业实行公私合营，安岳县仅余下饮食业个体户76户继续私营，而资阳县则将个体饮食店421户纳入组织。1958年，资阳地区成立大型国营、合作饮食店，如安岳县城新建了第一个国营餐厅——新安餐厅，由此，餐饮网点减少，菜肴趋向大众化。1966年~1976年，资阳地区城乡合营食店进入国营、合作商业后，分别由各县饮食服务公司和各区供销社饮食专业店统一安排，并增扩店或门市，分工经营各项业务，同时，各食店取消雅座，推广单锅炒菜等服务项目，顾客就餐自买牌子、自找桌凳、自取饭菜等。到1978年改革开放后，国营和集体饮食店等陆续实行经济承包责任制，个体户经营饮食逐步增多，恢复传统名菜点、品种增加，设立雅座，承包酒席，档次逐步提升，服务态度和质量逐步改善。仅以资阳县为例，1979年，全县饮食业有153户，从业人员1 275人。到1985年底，全县饮食业发展到1 077户。餐饮业的恢复与发展使得民众饮食生活水平逐步提高，一些名小吃也受到民众热情追捧，如郭饺子、钟焦粑、龙凉粉等。

From the founding of the People's Republic of China to the initial stage of Reform and Opening Up, the development of Ziyang cuisine still made great progress despite some twists and turns. In agriculture, Ziyang vigorously developed agriculture and promoted production, and the output of grain, oil, sugar cane and other crops gained a certain growth. For example, in the planting of economic crops, the planting area was about 149,300 mu in 1949, and was expanded to 537,400 mu in 1985. Moreover, Ziyang Corn Seed Stock Station and many other state-owned farms were established to promote the development of agriculture and provide rich and varied food materials for Ziyang cuisine. In the food and beverage manufacturing industry, Ziyang also had a great development. For example, Ziyang Sugar Factory was established in 1957, and Nantuanhe Sugar Factory was established in 1958. By 1985, according to the statistics of industry composition of total industrial output value, the output value of food manufacturing industry accounted for 21.38%, ranking the first, with 83 enterprises and output value of 80.49 million yuan. There were 32 beverage manufacturing companies with output value of 9.7 million yuan, accounting for 2.58%. The establishment and development of these enterprises promoted the transformation of Ziyang cuisine from individual handicraft industry to industrial production. In the catering industry, there were some development and transformation in Ziyang. According to the records of *Ziyang Annals* and *Anyue Annals*, in 1955, Anyue County had 1,301 private catering industry households, an increase of 66.4% compared to 1948. In 1956, the entire industry implemented the public-private partnerships; Anyue only remained 76 self-employed catering enterprises continuing to be privately owned, while Ziyang integrated 421 individual restaurants into the organization. In 1958, Ziyang area founded large state-owned restaurants and cooperative restaurants, such as Anyue opened the first state-owned restaurant—Xin'an Restaurant in the county. As a result, catering outlets reduced, dishes tended to be popularized. From 1966 to 1976, after the joint urban and rural restaurants of Ziyang became state-owned, cooperative businesses, the stores were arranged in a unified way by the catering service companies of each county and catering professional shops Supply and Marketing Cooperatives of each district. The shops or stores were expanded and different businesses were operated separately. At the same time, the restaurants canceled the private rooms, and promoted services like single pot cooking. Customers should buy the meal card, look for a table and stool, and take the food by themselves when dining. After the Reform and Opening Up in 1978, state-owned and collective eateries began to implement the economic responsibility contract system, and the self-employed gradually increased their catering diet, resumed the traditional famous dishes and enriched the varieties, set up private rooms, contracted banquets, gradually updated the level, and improved the service attitude and quality step by step. Take Ziyang County as an example, in 1979, there were 153 catering households and 1,275 employees in the county; while by the end of 1985, the number of the catering households in the county was up to 1,077. The recovery and development of the catering industry had gradually improved people's eating and living standards. Some famous snacks, such as Guo's Dumplings, Zhong's Crispy Pie, Long's Pea Jelly, etc., have also been warmly pursued by the public.

进入20世纪80年代中期尤其是21世纪以后，资阳美食在各级政府的支持和推动、餐饮从业人员的共同努力下逐渐走向高速发展阶段，呈现出以下5个显著特征。

Since the mid-1980s, especially after the 21st century, Ziyang cuisine has gradually entered the high-speed development stage with the support and promotion of governments at all levels and the joint efforts of catering practitioners, showing the following five significant features.

1 BOOMING FOOD INDUSTRY　美食产业繁荣兴旺

资阳文化与旅游资源优势突出，至2017年末，资阳市有国家A级景区9个，其中AAAA级景区2个，AAA级景区4个。同时，资阳市是四川省同时连接成渝"双核"的区域性中心城市，明显的区位优势将资阳市融入成都半小时经济圈、重庆一小时经济圈，大大增强了对外交通的高效衔接和快速通达性，助力资阳市成为成渝城市群最大的次级交通枢纽。2017年，资阳市全年实现旅游收入162亿元，接待国内外游客2 400万人次，极大地促进了餐饮市场的繁荣兴旺。

Ziyang has prominent advantages in cultural and tourism resources. By the end of 2017, Ziyang had 9 national A-level scenic spots, including two AAAA scenic spots and four AAA scenic spots. At the same time, Ziyang City is a regional central city in Sichuan Province connecting Chengdu-Chongqing "double core". The obvious location advantages integrate Ziyang into Chengdu half-hour economic circle and Chongqing one-hour economic circle, greatly enhance the efficient connection and rapid accessibility of external traffic, and help Ziyang become the largest secondary transportation hub of Chengdu-Chongqing urban agglomeration. In 2017, Ziyang achieved annual tourism revenue of 16.2 billion yuan and received 24 million domestic and foreign tourists, greatly promoting the prosperity of the catering market.

近年来，资阳美食产业发展迅速。据《资阳市年鉴》统计，2016年，全市餐饮收入69.2亿元，同比增长12.5%。其中，雁江区实现餐饮收入25.98亿元，增长14.3%；安岳县餐饮收入27.81亿元，同比增长11.9%；乐至县餐饮收入15.7亿元，同比增长10.4%。2017年，全市餐饮、零售等民生领域的投资同比增长62.3%，餐饮业营业额增长17.8%。其中，雁江区餐饮收入29.92亿元，同比增长14.9%；安岳县餐饮收入30.65亿元，同比增长12.4%；乐至县餐饮收入19.5亿元，同比增长17.2%。资阳市餐饮业态较为齐全，各业态都有众多名店，包括金迪大酒店、美食美味大酒店、小贝壳酒店等一大批知名餐饮企业；打造了苌弘文化宴、柠檬风味宴、长寿文化宴等一批主题宴会，涌现了刘官银、秦照明、肖述明等将传统文化融入美食之中的引领者、传播者，为弘扬资阳美食文化奠定了良好基础。

In recent years, the food industry in Ziyang has developed rapidly. According to the statistics of *Ziyang Yearbook*, in 2016, the catering revenue of the city was 6.92 billion yuan, with year-on-year growth of 12.5%. Among them, Yanjiang District achieved catering revenue of 2.598 billion yuan, with growth of 14.3%; the catering revenue in Anyue County reached 2.781 billion yuan, with year-on-year growth of 11.9%; the catering revenue in Lezhi County reached 1.57 billion yuan, with year-on-year growth of 10.4%. In 2017, the year-on-year growth of the city's investment in catering, retail and other livelihood areas was 62.3%, and the growth of turnover in catering industry was 17.8%. Among them, the catering revenue of Yanjiang District was 2.992 billion yuan, with year-on-year growth of 14.9%; the catering revenue in Anyue County reached 3.065 billion yuan, with year-on-year growth of 12.4%; the catering revenue in Lezhi County reached 1.95 billion yuan, with year-on-year growth of 17.2%. The catering business of Ziyang City is relatively complete, each business has numerous famous stores, including Jindi Hotel, Gourmet Restaurant, Little Shell Hotel and a large number of well-known catering enterprises; thus Changhong Cultural Feast, Lemon Flavor Banquet, Longevity Cultural Banquet and other theme banquets have been created, Liu Guanyin, Qin Zhaoming, Xiao Shuming and other pioneers and disseminators who integrated traditional culture into food have been sprung up, laying a good foundation for the promotion of food culture in Ziyang.

2 BASICALLY FORMED FOOD INDUSTRY CHAIN SYSTEM | 美食产业链体系基本形成

资阳美食市场的繁荣，极大地带动了与之紧密相连的原辅料、调料生产加工的兴旺。如今，资阳市形成了集食材的种植养殖、精深加工、经营销售、物流配送等于一体的美食产业集群，基本构建起了从田间到餐桌的美食全产业链体系。

The prosperity of the food market in Ziyang greatly promotes the flourishing of the production and processing of the closely relevant raw and auxiliary materials and seasoning. Nowadays, Ziyang City has formed a food industry cluster integrating the planting and cultivation of food materials, food intensive processing, operation and sales, and logistics and distribution, basically constructing the whole food industry chain system from field to table.

在食材的种植和养殖方面，资阳市不仅拥有长江中上游最大的早熟柑橘基地、全省乃至全国柠檬单一品种规模最大的基地，而且大力发展绿色蔬菜产业，重点打造了以雁江区、乐至县为主的100万亩绿色蔬菜发展区，建成李家坝、晏家坝、丹东路、祥符镇等四大万亩蔬菜基地。在食品生产加工业方面，资阳市近年来也发展迅猛。至2017年，全市已初步建成雁江的中和、安岳的龙台、乐至的中天3个农产品加工园区，规模以上食品工业企业96户，实现产值296.96亿元，占规模工业总产值的26.6%。资阳市涌现出了一大批著名的龙头食品企业，不仅有以当地农产品红薯、柠檬、青花椒、肉类等为原料的本土食品加工企业，如安岳薯霸食品有限公司、华通柠檬有限公司和乐至帅青花椒开发有限公司、乐至县金锄头粮油有限公司，以及雁江区的临江寺味业有限公司、永鑫肉类食品有限公司等，也有外来的著名食品企业在此建厂生产，如四川加多宝饮料有限公司、百威（四川）啤酒有限公司等。在物流配送方面，资阳市大力推进电子商务与现代流通体系相结合的发展模式，安岳县和乐至县已成为国家级电子商务进农村综合示范县，雁江区为省级电子商务进农村综合示范区。随着集食材种植养殖、精深加工、经营销售、物流配送等于一体的美食产业集群和产业链的不断发展、完善，将促进资阳餐饮产业规模的不断扩大和餐饮企业实力不断增强。

In terms of the planting and cultivation of food materials, Ziyang City not only has the largest precocious citrus base in the upper and middle reaches of the Yangtze River and the largest single variety of lemon base in the province or even in the whole country, but also vigorously develops the green vegetable industry, mainly building the Yanjiang District and Lezhi County based Green Vegetable Development Zone of 1 million mu, and setting up 4 Ten Thousand Mu Vegetable Base of Li's Ba, Yan's Ba, Dandong Road and Xiangfu Town. In the food production and processing industry, Ziyang City has also developed rapidly in recent years. By 2017, the city had initially built three agricultural products processing parks, namely Zhonghe of Yanjiang, Longtai of Anyue and Zhongtian of Lezhi, with 96 food industrial enterprise above designated size and an output value of 29.696 billion yuan, accounting for 26.6% of the total output value of the scale industry. A large number of famous leading food companies, including not only the local food processing enterprises such as Anyue Shuba Food Co., Ltd., Huatong Lemon Co., Ltd., Lezhi Shuai Green Sichuan Pepper Development Co., Ltd., Lezhi Gold Hoe Grain and Oil Co., Ltd., and Linjiangsi Temple Taste Industry Co., Ltd. and Yongxin Meat Products Co., Ltd. of Yanjiang District, etc. which took sweet potato, lemon, green Sichuan pepper, meat and other local agricultural products as raw materials; but also the ecdemic famous food enterprises such as Sichuan JDB Beverage Co., Ltd., Budweiser (Sichuan) Beer Co., Ltd., etc. have emerged in Ziyang and built factories and manufacture here. In terms of logistics and distribution, Ziyang City vigorously promotes the development model of combining e-commerce with modern circulation system. Anyue County and Lezhi County have become national comprehensive demonstration counties of e-commerce in rural areas, and Yanjiang District has been provincial comprehensive demonstration area of e-commerce in rural areas. With the continuous development and improvement of food industry clusters and industrial chains integrating the planting and cultivation of food materials, food intensive processing, operation and sales, and logistics and distribution, the scale of Ziyang catering industry will be continuously expanded and the strength of catering enterprises will be constantly enhanced.

安岳魅力柠海　Glamorous Lemon Sea in Anyue County

3 DISTINCTIVE AND COLORFUL ZIYANG CUISINE | 资阳美食特色鲜明且异彩纷呈

资阳美食以资阳的地理、物产、经济、习俗等为依托，逐渐具有了鲜明而突出的特色，主要表现在3个方面：

Relying on the geography, products, economy and customs of Ziyang, Ziyang cuisine has gradually acquired distinct and prominent characteristics, mainly manifested in three aspects.

资阳美食的食材种类丰富、品质优良。资阳市属亚热带湿润季风气候区，四季分明，热量充足，雨量充沛，江河渠系纵横交错，平坝、丘陵、山地起伏相连，土壤以棕紫泥土为主，钾元素含量高，自然资源极为丰富，适宜柠檬、柑橘、山羊、生猪、蚕桑等川中丘陵地区特色农林牧渔各业生产。资阳市通过积极引进新品种、新技术，提高单位面积产量，形成了"一带多片"优质粮油基地布局，粮食自给率

资阳特色餐饮　Characteristic Ziyang Food

达到95%以上；大力发展柑橘、柠檬等精品水果及绿色蔬菜产业，种植业结构不断优化，效益显著提升；全市通过大力发展以生猪为支柱和以山羊为优势的外向型畜牧业，走出了一条丘陵地区发展现代畜牧业的新路子。3个县（区）均是国家商品粮基地县，并被认定为全省现代畜牧业重点县。其中，安岳柠檬产量、规模和市场占有率均占全国80%以上；乐至蚕桑综合产值居全省第二位；雁江柑橘产量居全省第二位。此外，全市花椒、核桃、白乌鱼、泥鳅等产业也初具特色和规模。经过多年发展，资阳市已形成优质粮油、畜牧、水果、蔬菜等支柱产业，为资阳美食发展提供了种类丰富、品质优良的食材。

Firstly, Ziyang cuisine has rich variety and high quality of food materials. Ziyang City is a subtropical humid monsoon climate zone with distinct seasons, sufficient heat, and abundant rainfall. The river systems are crisscrossed, and plain, hills and mountains are linked together. The land is mainly brown purple soil which is rich in potassium. So the natural resources here are abundant. It is suitable for the production of lemon, citrus, goat, pig, silkworm and other hilly region characteristic agriculture, forestry, animal husbandry and fishery products in center Sichuan. By actively introducing new varieties and technologies to improve the yield per unit area, Ziyang City has formed a distribution of high-quality grain and oil base of "one belt and several areas", and the grain self-sufficiency rate reaches more than 95%. Meanwhile, Ziyang vigorously develops the citrus, lemon and other fine fruit and green vegetable industries, optimizes the planting structure, and significantly improves the benefits. Through developing the export-oriented animal husbandry with pig as the pillar and goat as the advantage, the whole city has found a new way to develop modern animal husbandry in hilly areas. All of the three counties (districts) are National Commodity Grain Base Counties, and are identified as the provincial key counties of modern animal husbandry. Among them, the production, scale and market share of Anyue lemon account for more than 80%; the comprehensive production of Lezhi silkworm ranks second in the province; the production of Yanjiang citrus holds the second place in the province. In addition, the industries of Sichuan pepper, walnut, white snakehead, loach, etc. has formed certain characteristics and scale. After years of development, Ziyang City has developed high quality grain and oil, livestock, fruit, vegetable and other pillar industries, providing a rich variety and good quality of food materials for the great food development.

资阳美食的风味源远流长、名品较多。史前时期的资阳人即已在资阳地区开始饮食生活，经过漫长的历史发展，资阳美食形成了独特的风味。在清代末年成熟定型的川菜体系中，资阳菜占据一席之地。如今，资阳菜在风味上不仅注重麻辣多变、味道浓厚，而且注重咸鲜醇香、略带甘甜。资阳河流纵横，水产丰富，人们用当地出产的青花椒、紫竹姜、辣椒等调味料制作出麻辣味厚的众多河鲜菜品，香辣小龙虾、苌弘鲶鱼、藿香乌鱼花等即是其中的著名品种。资阳自古便是重要的蔗糖生产地，人们不仅喜食甜食，也善烹甜食，天长日久，相沿成习，制作出了乐至麻饼、糍粑小龙虾等味道甘甜的菜品。近年来，在资阳菜点中，获得国家级、省级、市级名菜点称号的超过百余种。

Secondly, Ziyang cuisine has a long history and lots of famous products. In prehistoric times, people in Ziyang began to eat and drink in Ziyang area. After a long history of development, Ziyang cuisine formed a unique flavor. During the late Qing Dynasty, in mature and fixed Sichuan cuisine system, Ziyang cuisine had a special place. Nowadays, Ziyang dishes not only pay attention to spicy and changeable flavor, strong taste, but also emphasize on being salty, fresh, mellow, and slightly sweet. This city is river-crisscrossed and rich in aquatic products, and people use local seasonings such as green Sichuan pepper, Zizhu ginger and chili to make fresh river dishes with a spicy flavor. Famous varieties such as Crayfish in Chili Sauce, Changhong Catfish, and Deep Fried Snakehead with Huoxiang Herb are some of them. Ziyang has been an important sugar producing area since ancient times. People not only like sweet food, but also like to cook sweet food. Over time, they become accustomed to making sweet dishes such as Honey Lotus Flower, Lezhi Sesames Pastry and Fried Crayfish with Brown Sugar and Glutinous Rice Cake. In recent years, more than 100 kinds of national, provincial and municipal famous dishes have been produced in Ziyang.

资阳美食的制作技艺精湛。资阳自古以来名人辈出，其中有不少名厨大师，如中国国宝级川菜大师陈松如，就曾为毛泽东、朱德、周恩来、邓小平等党和国家领导人以及外国首脑烹调川菜珍肴，受到高度赞誉。资阳美食制作技艺不断传承创新，一直延续至今。目前，资阳美食的相关制作技艺已列入省级非物质文化遗产名录1项，即临江寺豆瓣传统工艺；列入市级非物质文化遗产名录7项，包括两节山老酒传统酿造技艺、苌弘鲶鱼烹调技艺、安岳米卷制作技艺、乐意窖藏酒制作技艺、乐至仙荷藕粉制作技艺、乐至外婆坛子肉制作技艺、乐至烤肉制作技艺。此外，在资阳民间家庭和餐饮企业中还有大量精湛的菜点传统制作技艺在传承发展，如切割纤细的云白豆腐丝、煎制得薄如蝉翼的桑叶薄脆等菜点制作，丰富着民众饮食生活，承载着资阳人家乡的味道。

Thirdly, the cooking skills of Ziyang cuisine are exquisite. Celebrities have come out in large numbers in Ziyang since ancient times. Among them, there are a lot of famous master chefs, such as China national treasure of Sichuan cuisine master Chen Songru, who cooked Sichuan dishes for Mao Zedong, Zhu De, Zhou Enlai, Deng Xiaoping and other Party and state leaders and foreign leaders, and was highly acclaimed. Ziyang food production skills continue to be inherited and innovated till now. At present, one item of the related Ziyang food production skills has been included in provincial intangible cultural heritage list, namely the traditional handicraft of Linjiangsi chili bean paste; 7 items have been included in municipal intangible cultural heritage list, covering the traditional brewing craft of Liangjieshan Aged Liquor, the cooking skills of Changhong Catfish, the production skills of Anyue rice rolls, the production techniques of Leyi Cellar Liquor, the production craft of Lezhi Xianhe Lotus Root Porridge, the cooking techniques of Lezhi Grandma's Tanzi Pork, and the production techniques of Lezhi Pork BBQ. In addition, in common families and catering businesses of Ziyang, a large number of exquisite traditional production techniques of dishes are inherited and developed, such as the skills of finely cut shredded White Cloud Doufu slivers, Crispy Mulberry Leaf Pancake as thin as cicada's wing of mulberry leaves, etc., which enrich people's catering life, and bear the hometown flavor of Ziyang people.

4 RESULTS OF CATERING TALENT TEAM CONSTRUCTION AND BRAND CREATION | 餐饮人才队伍建设与品牌打造成效显现

资阳市长期以来注重餐饮人才队伍建设。其中，资阳饮食服务公司有着独特的贡献。据公司资料显示，1999年，该公司已有特二级烹调师6名，特三级烹调师8名，一级烹调师44名，二级烹调师49名；特二级面点师1名，一级面点师8名，二级面点师13名，先后为北京、重庆、南京、上海等地输出300余名餐饮人才，拥有培训基地，实作、鉴定场地2 500平方米，并获批成立了"资阳地区饮食服务行业国家职业技能鉴定所"。该公司一直延续至今，成为当代资阳餐饮业发展历史的见证者和资阳美食人才的重要摇篮。此外，资阳市的相关学校、餐饮行业协会也积极开展餐饮人才培养与能力提升工作。如资阳市雁江区职业技术学校、安岳第一职业技术学校、乐至县高级职业中学等，都在其旅游服务专业中设有餐饮服务等相关课程；资阳市及所属各县（区）餐饮协会也通过烹饪比赛、餐饮技能和管理培训等形式提升餐饮从业人员的技能和管理水平。

Ziyang City has paid attention to the construction of catering talent team for a long time. Thereinto, Ziyang Catering Service Company has a unique contribution. According to the company information, in 1999, the company had 6 super-class II chefs, 8 super-class III chefs, 44 first-level chefs and 49 second-level chefs; 1 super-class II baker, 8 super-class II baker, 13 second-level. It has exported more than 300 catering talents to Beijing, Chongqing, Nanjing, Shanghai and other places. It has a training base with a field of 2500 square meters for practice and identification, and has been approved to set up "National Vocational Skill Identification Institute of Ziyang Catering Service

Industry". Till now, it has become the witness of the development history of modern Ziyang catering industry and the important cradle of talents in Ziyang cuisine. In addition, relevant schools and catering industry associations in Ziyang also actively carry out catering talent training and capacity improvement activities. For example, Ziyang Yanjiang District Vocational and Technical School, Anyue No.1 Vocational and Technical School, Lezhi Senior Vocational High School, etc. all have catering service and other related courses in their tourism service majors. Catering associations in Ziyang City and the subordinate counties (districts) improve the skills and management level of catering practitioners via cooking competitions, catering skills and management training.

资阳市十分注重美食品牌的打造。2016年，资阳市启动"资味"区域公用品牌创建，成立"资味"区域公用品牌发展协会，现有会员单位46家，申请"资味"商标注册11个类别中的6个类别已通过国家商标局审核。截至2017年底，全市有中国驰名商标4个、四川著名商标10个、四川名牌5个、中华老字号1个、四川老字号3个，其中，安岳柠檬区域品牌价值达173.61亿元，位居全国农产品品牌价值第24位。全市通过无公害种植业农产品产地整体认定，认定面积307.5万亩；获认证的无公害畜禽产品产地20个，绿色食品原料生产基地3个、面积46.6万亩。全市认证"三品一标"农产品191个，其中，无公害农产品131个，绿色食品25个，有机食品28个，国家地理标志保护产品7个。近年来，资阳市及所属各县（区）还着力打造了多个美食节会品牌，如"资味"美食文化节、资阳市乡村旅游美食节、苌弘美食节、乐至烧烤节、安岳柠檬风味美食节、中和小龙虾节等，以此进一步发掘资阳美食和饮食文化，着力打造资阳本地餐饮品牌，营造餐饮消费氛围，促进社会消费。

Ziyang City also attaches great importance to the creation of food brands. In 2016, Ziyang City launched the creation of "Zi Wei" regional public brand and established "Zi Wei" Regional Public Brand Development Association. Currently, there are 46 member units. Six of the eleven categories of "Zi Wei" trademark registration applications have been approved by the State Trademark Bureau. By the end of 2017, there were 4 well-known trademarks of China, 10 famous trademarks of Sichuan, 5 Sichuan famous brands, 1 China time-honored brand and 3 Sichuan time-honored brands. Among them, the brand value of Anyue Lemon reached 17.361 billion yuan, ranking 24th in the brand value of agricultural products in China. The whole area of pollution-free planting agricultural products has been identified as 3.075 million mu. It has 20 certified producing areas of pollution-free livestock and poultry products, 3 production bases of green food raw materials with an area of 466,000 mu. There are 191 agricultural products with "Three Products and One Indication" certified by the city, including 131 pollution-free agricultural products, 25 green foods, 28 organic foods and 7 national geographical indication protected products. In recent years, Ziyang and the subordinate counties (districts) also strive to build lots of food festival brands, such as "Zi Wei" Food Culture Festival, Ziyang Rural Tourism Food Festival, Changhong Food Festival, Lezhi Barbecue Festival, Anyue Lemon Flavor Food Festival, and Zhonghe Crayfish Festival and so on, so as to further explore Ziyang food and food culture, strive to build great local catering brand, create the atmosphere of catering consumption, and promote social consumption.

5　ORGANIC COMBINATION OF GOVERNMENT ATTENTION AND ASSOCIATION PROMOTION　政府重视与协会助推有机结合

长期以来，资阳各级政府十分重视美食产业发展和饮食文化的挖掘与传承。随着资阳大力实施产业强市战略，构建以新兴产业为先导、先进制造业为引领、现代服务业为支撑、现代农业为基础的产业体系，为美食产业的转型发展提供了巨大空间，取得了许多成效。如在旅游餐饮方面，至2017年，资阳已有旅游特色餐饮单位27家，乡村旅游特色乡镇6个、村8个，星级农家乐/乡村旅店36家。2019年，资阳市为成功举办"第二届世界川菜大会"，进一步推动美食产业发展，在资阳市商务局成立了"川菜办公室"，负责统筹协调相关工作。此外，各级政府十分重视食品安全工作。2017年，资阳市启动四川省食品药品安全示范市、州的创建工作，全市14家规模以上食品生产企业通过质量体系认证，全市4 618家持证餐饮服务单位完成量化评级工作。各级政府还不断加强"三小"（食品小作坊、小经营店及小摊贩）管理，出台认定标准和指导意见，全面推行备案登记管理；在全省率先开展"互联网+"食品药品网格化监管试点工作，实现市、县、乡、村全面覆盖；率先制定《农村群体性聚餐监管办法》及操作指南，组建"乡厨协会"，实行"坝坝宴"监管九统一，至2017年，全市纳入统计规范管理的乡厨队伍有951个。

For a long time, all levels of government in Ziyang attach great importance to the development of food industry and the excavation and inheritance of food culture. With Ziyang vigorously implementing the strategy of strengthening the city by industry, the establishment of an industrial system with emerging industries as the precursor, advanced manufacturing as the guide, modern service industry as the support, and modern agriculture as the basis, provides a huge space for the transformation and development of food industry, and has achieved a lot of results. For example, in terms of tourism catering, by 2017, there were 27 tourism characteristic catering units, 6 rural tourism characteristic towns and 8 villages, and 36 star-rated farms/village hotels. In 2019, in order to successfully hold The 2nd World Sichuan Cuisine Conference and further promote the development of food industry, Ziyang Municipal Bureau of Commerce established "Sichuan Cuisine Office", responsible for coordinating relevant work. In addition, governments at all levels attach great importance to food safety work. In 2017, Ziyang launched the establishment of Sichuan food and drug safety demonstration cities/states; 14 food production enterprises above

designated size were certified by quality system, and 4,618 licensed catering service units completed quantitative rating. Governments at all levels also constantly strengthen the "Three Small" (small food workshops, small business shop and small stalls) management, introduce standards and guidelines, and comprehensively implement the record and registration management; take the lead in carrying out "Internet +" food and drug grid supervision pilot work in the province, and achieve full coverage of cities, counties, townships and villages; formulate *Regulatory Measures for Group Dining in Rural Areas* and operation guidelines first, set up township chef association, and carry out "Ba Ba Feast" nine unification in regulation. Till 2017, there were 951 township chef teams under the statistics and standard management in Ziyang.

在资阳各级政府及有关部门高度重视和支持下，资阳各级餐饮协会等相关社会团体也充分发挥作用，搭建政府与餐饮企业间的纽带和平台，及时传递政策信息，挖掘传承资阳美食文化，定期组织举办美食节、烹饪比赛、对外交流等活动，开展餐饮从业人员培训工作，提升从业人员素质。如2015年，第一届资阳美食节由资阳市餐饮协会等主办，评选出资阳名菜60个、资阳名小吃16个、资阳名店12个、资阳名厨23个、资阳明星服务员12个、资阳创意菜品10个。2018年，资阳市第二届"资味"美食文化节由资阳市商务局主办，资阳市餐饮协会等承办，也评选出了资阳名菜、名店、名厨、名料、名小吃、优秀服务员、优秀职业经理人和餐饮发展特别贡献人物，极大地推动了资阳美食品牌的打造。此外，资阳市餐饮协会还先后编撰了《资阳人家——美食美味》《苌弘演义》等相关书籍。

Due to the attention and support of governments at all levels and relevant departments, the catering associations at all levels and relevant social groups also full play their roles to build the link and platform between the government and catering enterprises, timely deliver policy and information, excavate and inherit Ziyang food culture, hold food festivals, cooking contests, external exchange and other activities regularly, carry out training for catering practitioners, and improve staff quality. For example, in 2015, the first Ziyang Food Festival was majorly hosted by the

Ziyang Catering Association, which selected 60 Ziyang famous dishes, 16 Ziyang famous snacks, 12 Ziyang famous restaurants, 23 Ziyang famous chefs, 12 Ziyang star waiters, and 10 Ziyang creative dishes. In 2018, The 2nd Ziyang "Zi Wei" Food Culture Festival was sponsored by Ziyang Bureau of Commerce and undertaken by Ziyang Catering Association, etc. It also selected Ziyang famous dishes, famous restaurants, famous chefs, famous materials, famous snacks, excellent waiters, excellent professional managers and special contributors to the development of catering, which greatly promoted the establishment of Ziyang food brands. In addition, Ziyang Catering Association has successively compiled *Ziyang Family—Delicious Food*, *Romance of Chang Hong*, and other relevant books.

总之，资阳美食历史悠久，发展至今，依托丰富的自然资源，在政府引导、协会助推、市场主导、企业运作、立足民间等发展原则指导下，不仅拥有了鲜明特色、异彩纷呈的美食，而且形成了美食全产业链体系，美食产业繁荣兴旺，餐饮人才队伍建设与品牌打造取得显著成效。2019年9月，"第二届世界川菜大会"将在资阳召开，这是资阳美食产业发展进程中的一个新里程碑。资阳市将以此为契机，不断挖掘、传承和弘扬美食文化，不断创新、完善美食产业体系，促进一、二、三产业联动和文旅融合互动，促进资阳美食产业转型发展，进一步地推动资阳美食走遍全国、走向世界，提升其国际知名度和文化软实力。

All in all, Ziyang cuisine has a long history. Relying on the abundant natural resources, under the direction of government guiding, association helping, market leading, enterprise operation, base on the folk and other development principles, it not only has distinct characteristic, but also forms a food chain system. The food industry is flourishing, and remarkable achievements have been made in the construction of catering talent team and brand creation. In September, 2019, "The 2nd World Sichuan Cuisine Conference" will be held in Ziyang, which will be a new milestone in the development of the food industry in Ziyang. Ziyang will take this opportunity to continuously explore, inherit and carry forward the food culture, constantly innovate and improve the food industry system, promote the linkage of the first, second and third industries and the integration and interaction of cultural tourism, accelerate the transformation and development of food industry in Ziyang, and further spread Ziyang cuisine throughout the country and to the world, and raise its international reputation and culture soft power.

Chapter Two

Geographical Environment and Famous and Special Food Materials of Ziyang Cuisine

As the saying goes, the unique features of a local environment always give special characteristics to its inhabitants. Geographical environment is the precondition for the production of food materials and the major influencing factor of people's dietary life. Ziyang City has superior natural conditions, the terrain is diverse, the rivers are crisscross, the climate is warm, and the scenery is pleasant. In this beautiful and rich land, there are various kinds of food materials, including lots of famous and special food materials, which nourish generations of Ziyang people to survive and reproduce here, bring up the characteristic catering customs of Ziyang, and serve as the foundation of Ziyang cuisine.

第二篇 资阳

美食地理环境与名特食材

俗话说：一方水土养一方人。地理环境是食材出产的先决条件，是民众饮食生活的主要影响因素。资阳市自然条件非常优越，境内地形多样，河流纵横，气候温润，景色怡人。在这片美丽而富饶的土地上，生长着各种各样的食材，其中不乏名特食材，它们滋养着代代资阳人在此生存、繁衍，造就了资阳特色饮食民俗，是资阳美食的主要基础保障。

GEOGRAPHICAL ENVIRONMENT
地理环境

位置与气候 | LOCATION AND CLIMATE | 1

资阳，四川省地级市。位于东经104°21'～105°27'，北纬29°15'～30°17'，在成都和重庆两大城市的中间。北靠成都（相距88公里），南连内江，东接重庆（相距257公里）、遂宁，西邻眉山。现总辖雁江区、安岳县、乐至县，总面积约5 757平方公里，资阳市人民政府驻雁江区广场路。资阳是同时连接成渝"双核"的四川省区域性中心城市，交通非常便利，成渝高铁、成渝铁路、成都地铁18号线和成渝、渝蓉、遂资眉等高速路网体系四通八达，资阳主城区距成都天府国际机场仅18公里。随着成资渝高速公路、蓉昆高铁的加快建设，"多线接成渝，市域大畅通"的综合交通格局正在形成。

资阳市一区两县规划示意图

Planning Sketch of One District and Two Counties of Ziyang City

Ziyang is a prefecture-level city in Sichuan Province. It is located at 104°21'~105°27' east longitude and 29°15'~30°17' north latitude, in the middle of Chengdu and Chongqing. It is adjacent to Chengdu in the north (88 kilometers apart), Neijiang in the south, Chongqing (257 kilometers apart) and Suining in the east, Meishan in the west. It has jurisdiction over Yanjiang District, Anyue County and Lezhi County, with the total area of about 5,757 square kilometers. The municipal people's government is stationed on Square Road, Yanjiang District. Ziyang is a regional central city in Sichuan Province connecting Chengdu-Chongqing "double core". The transportation is very convenient as Chengdu-Chongqing High-Speed Rail, Chengdu-Chongqing Railway, Chengdu Metro Line 18 and Chengdu-Chongqing, Yurong, Suizimei Expressways and other road network system extend in all directions. Ziyang main urban area is only 18 kilometers away from Chengdu Tianfu International Airport. With the acceleration of the construction of Chengziyu Expressway and Rongkun High-Speed Rail, the comprehensive traffic pattern of "multiple lines connecting Chengdu and Chongqing and cities with great unimpeded traffic" is taking shape.

资阳属亚热带湿润季风气候区，气候温和湿润，四季分明，终年碧绿，常年平均温度在17℃左右，年平均降雨量1 100毫米，年日照时数1 300小时，年平均无霜期长达300天。全年云雾多而日照少，空气湿度大而昼夜温差小。最热月8月，平均气温26.5℃左右；最冷月1月，平均气温6.5℃左右；极端最高气温40.2℃；极端最低气温–5.4℃。就盆地气候而言，资阳市还具有南–北、东–西气候过渡带的特点，境内的西北部龙泉山海拔高度600~1 000米，比丘陵区高出300~500米，其产生的空气下沉增温减湿作用对资阳气候有一定影响。

Ziyang is a subtropical humid monsoon climate zone with mild and humid climate and distinct seasons, and is green all year round. The perennial average temperature is about 17℃, the average annual rainfall is 1,100 mm, the annual sunshine hours are 1,300 hours, and the average annual frost-free period is up to 300 days. There are more clouds and less sunshine throughout the year. The air humidity is high and the temperature difference between day and night is small. In August, the hottest month, the average temperature is about 26.5℃. In January, the coldest month, the average temperature is about 6.5℃. Extreme maximum temperature is 40.2℃, while extreme minimum temperature is -5.4℃. In terms of basin climate, Ziyang City also has the characteristics of the south-north and east-west climate transition zone. The altitude of Longquan Mountain in the northwest of Ziyang is 600~1,000 meters, which is 300~500 meters higher than that of hilly area. The resulting air subsidence effect of increasing temperature and reducing humidity has a certain impact on Ziyang's climate.

水文与地貌　HYDROLOGY AND GEOMORPHOLOGY　2

　　资阳市境内水资源以河川径流为主，有2/3地域处于沱江与涪江的两江分水岭，是川中径流低值区，川东伏旱与川西夏旱交错。全市河流分属二个水资源三级区，水域面积30多平方公里，流域面积达2 000多平方公里；共有沱江与涪江的两江支流（中、小河流）110条，其中，流域面积大于100平方公里的河流就有11条，流域面积50～100平方公里的小河8条。此外，还有短小溪流40余条，几乎都发源于丘陵，河床平、缓、宽，落差小、水流平缓、岸势开阔，是典型的丘陵地区水系网络。

　　The water resources in Ziyang City are mainly river runoff, 2/3 of which are located in the watershed of Tuojiang River and Fujiang River. It is the low runoff area in the middle of Sichuan. Summer droughts occur alternatively in East Sichuan and West Sichuan. The rivers in the city belong to two Three-level Water Resources Zones, with a water area of more than 30 square kilometers and a total drainage area of more than 2,000 square kilometers. There are 110 tributaries (medium and small rivers) of Tuojiang River and Fuljiang River, among which there are 11 rivers with each drainage area larger than 100 square kilometers and 8 rivers with each drainage area between 50 and 100 square kilometers. In addition, there are more than 40 small streams, almost all of which originate from hills. The riverbed is flat, slow and wide, with small drop, gentle water flow and open and broad river bank, which is a typical water system network in hilly areas.

　　资阳市的地貌形态大致分为三种类型，即低山、丘陵、河流冲积坝等。其中以丘陵为主，大约占总面积的90%以上，一般海拔在300～550米，河坝的最低点在夏家坝的琼江河出界处（海拔247米）。

　　The landform of Ziyang City can be roughly divided into three types, namely, low mountains, hills, river alluvial dams and so on. Among them, the major landform is hills, accounting for more than 90% of the total area, generally between 300 and 550 meters above sea level. The lowest point of the river dike is at the Qiongjiang River boundary of Xia's Ba (247 meters above sea level).

FAMOUS AND SPECIAL FOOD MATERIALS
名特食材

资阳温和湿润的气候、纵横交错的河流，以及多种类型的地貌等良好的自然地理环境，使得全市的动植物资源十分丰富，共有野生动物236种，野生植物有2 000多种，森林覆盖率达39.8%。得天独厚的地理环境也孕育出资阳非常丰富和优质的食材，为资阳美食文化的发展奠定了坚实的物质基础。

Ziyang's mild and humid climate, crisscrossed rivers and various types of landforms and other good natural geographical environment make the city rich in animal and plant resources. It has 236 species of wild animals and more than 2,000 species of wild plants, and the forest coverage rate reaches 39.8%. The favorable geographical environment also breeds rich and high-quality food materials in Ziyang, which lays a solid material foundation for the development of the food culture in Ziyang.

自古以来，资阳市作为四川省农作物主产区之一，农牧业基础好，特色农业优势明显，是天府粮仓、鱼米之乡。如今，资阳是"全国农业产业化工作先进市"，3个区、县均是国家商品粮基地县，安岳、乐至为中国肉类产量百强县，生猪、山羊、水产、水果、蚕桑、优质油料的产量均居四川省前列。资阳特色食材品种繁多，产量大且品质优，许多特色食材是国家地理标志保护产品并且成为四川乃至全国知名品牌，远销海内外。它们是资阳农村经济的重要组成部分，是农民脱贫致富的法宝，也是资阳美食文化的靓丽名片。由于资阳市特色食材数量繁多，本书限于篇幅，主要选择获得中国国家地理标志产品称号、列入非物质文化遗产保护名录，以及传统名特产品或种植面积与产量突出等极具代表性的名特色食材进行介绍。

Since ancient times, Ziyang has been one of the main crop producing areas in Sichuan Province, with good agricultural and animal husbandry foundation and obvious advantages in characteristic agriculture. It is the Tianfu granary and a land of fish and rice. Today, ziyang is a "National Advanced City in Agricultural Industrialization". The three districts and counties are all national commodity grain base counties. Anyue and Lezhi are China's top 100 counties in meat production. The output of pigs, goats, aquatic products, fruits, sericulture and high-quality oil plants all rank the top in Sichuan Province. Ziyang has a wide variety of specialty food materials with large output and excellent quality. Many special food materials are national geographic indication protection products and have become famous brands in Sichuan and even the whole country, selling well at home and abroad. They are important components of the rural economy of Ziyang, magic weapons for farmers to get rid of poverty and become better off and also beautiful the name cards of the food culture of Ziyang. Due to the large quantity of characteristic food materials in Ziyang and because of the limited space in this book, it mainly chooses the very representative famous and special food materials that have won the title of China's National Geographic Indication Product or have been included in the protection list of intangible cultural heritage, as well as traditional famous and special products or products with outstanding planting area and yield, etc.

1 GRAIN, VEGETABLES AND FRUITS | 粮食蔬果类

① 安岳红薯 | Anyue Sweet Potato

安岳红薯，又称为安岳红苕，是资阳市安岳县的著名特产。红薯营养丰富，含有多种人体需要的营养物质，包括蛋白质、糖、脂肪、磷、钙、铁和多种维生素等，对抗癌、预防心血管疾病等有独特作用，被誉为"天然绿色食品"和"最佳营养保健品"。

Anyue sweet potato is a famous specialty of Anyue County of Ziyang City. Sweet potato is rich in nutrition, containing a variety of nutrients needed by the human body, including protein, sugar, fat, phosphorus, calcium, iron and a variety of vitamins, etc. It has a unique role in anti-cancer, prevention of cardiovascular disease and other diseases, and is known as "natural green food" and "the best nutrition and health care product".

安岳红薯常年种植面积50余万亩，鲜薯产量70余万吨，年加工鲜薯35万吨以上，种植面积与产量均居全省各县（市）第一。全县红薯良种繁育基地乡镇14个、村129个、社1201个，良种种源面积达5万亩。2018年，安岳鲜薯产值约4亿元，销售收入3.3亿元。安岳红薯不仅直接销售，还进行精深加工。安岳县的红薯加工产业发达，现已成为西南地区最大的红薯粉条生产加工基地。全县从事红薯加工的农户达12万户、从业人员有30万余人；建有大型红薯加工企业2个、粉条专业市场4个，其中1个是西南地区最大的粉条专业市场。其红薯加工系列产品有精白淀粉、各种粉条、快餐粉丝等，总产量近5万吨，遍销全国各省、市。

The perennial planting area of sweet potato in Anyue is more than 500,000 mu, the output of fresh sweet potato is more than 700,000 tons, and the annual processing quantity of fresh sweet potato is more than 350,000 tons. The planting area and yield rank first in all counties (cities) of Sichuan. There are 14 sweet potato fine seed breeding base towns, 129 villages and 1,201 communities in the county, with a fine seed provenance area of 50,000 mu. In 2018, the output value of Anyue fresh sweet potatoes was about 400 million yuan and the sales revenue was 330 million yuan. Anyue sweet potatoes are not only sold directly, but also processed intensively. Since the sweet potato processing industry of Anyue is very developed, it now has become the largest sweet potato vermicelli production and processing base in southwest China. There are 120,000 peasant households and over 300,000 practitioners in the county engaged in sweet potato processing industry; there are 2 large sweet potato processing enterprises and 4 professional vermicelli markets, among which 1 is the largest professional vermicelli market in southwest China. The sweet potato processing series products include refined white starch, all kinds of vermicelli, fast food vermicelli, etc., with the total output of nearly 50,000 tons, selling all over the country.

② 周礼粉条 | Zhouli Vermicellii

周礼粉条，资阳市安岳县特产，由该县周礼镇所产，中国国家地理标志产品，有"素鱼翅"之称。早在清代道光年间，周礼镇就有了制作粉条的记载。周礼粉条采用先成型、再熟化的工艺，经过红薯取粉、淀粉沉淀、干燥打糊、漏粉、冻库屯条、干燥包装等多道工序制作而成，粉条色泽黄中带黑，有光泽，复水后呈半透明状，具有粗细均匀、细腻、柔软、不断条、不浑汤的特点。2011年12月26日，原国家质检总局批准对周礼粉条实施地理标志产品保护。如今，"周礼粉条"正处于全面现代化生产的转型期，年产值达5亿～6亿元。安岳县大力发展红薯优势产业，加快建设百亿产业园区，打造特色薯业小镇，力争将"周礼粉条"打造成中国知名品牌。

Zhouli vermicelli, a special product of Anyue County, Ziyang City, is produced in Zhouli Town. It is China's National Geographic Indication Product, known as "artificial Sharks fin". As early as in the Daoguang Period of the Qing Dynasty, there existed the records of Zhouli Town making vermicelli. Adopting the technology of first molding and then curing, Zhouli vermicelli is made through a variety of processes of crushing the sweet potatoes into powder, starch precipitation, making the dry powder into paste, griddling the paste into vermicelli, storing the vermicelli in freezer, packing after drying and so on. The vermicelli is yellow with blackish and shiny, and is translucent after rehydration, with the characteristics of being evenly thick, fine and smooth, soft, not easy to broken, and clear soup. On December 26, 2011, the former General Administration of Quality Supervision, Inspection and Quarantine of the People's Republic of China (AQSIQ) approved the implementation of geographical indication product protection for Zhouli vermicelli. Today, "Zhouli Vermicelli" is in the transition period of comprehensive modern production, with an annual output value of 500~600 million yuan. Anyue County will vigorously develop the advantage industry of sweet potato, accelerate the construction of ten billion industrial parks, build a characteristic sweet potato industry town, and strive to make "Zhouli Vermicelli" into a well-known brand in China.

周礼粉条在菜点制作中运用广泛，常用于凉拌、热炒、焖炖、火锅、煲汤等，是鱼香粉条、酸辣粉、凉拌粉丝、猪肉炖粉条等菜肴的理想食材。

Zhouli vermicelli is widely used in the production of dishes, and is often used in cold salad, hot frying, braising, hot pot, soup, etc. It is an ideal food material for Fish Flavor Vermicelli, Hot and Sour Vermicelli, Cold Vermicelli Salad, Braised Pork with Vermicelli and other dishes.

③ 安岳米卷 | Anyue Rice Roll

安岳米卷，资阳市安岳县特产。安岳米卷制作技艺于2009年被资阳市人民政府列入第二批市级非物质文化遗产代表性名录。

Anyue rice roll is a special product of Anyue County, Ziyang City. In 2009, the production techniques of Anyue rice roll were included in the representative list of the second batch of municipal intangible cultural heritage by Ziyang Municipal People's Government.

安岳米卷历史悠久，做工精细。相传清代末年，安岳人蒙吉安在云南学会米卷制作技术后，回到安岳县城独家开店，安岳米卷自此问世。安岳米卷选用当地的优质水源、优质大米，将大米浸泡后用石磨磨细、吊浆、摊皮，再上锅蒸制，晾凉后卷裹成筒即可，全部由手工完成。根据加入原料和辅料的不同，有白米卷、黄米卷、黑米卷、绿米卷等多个品种。各种米卷呈圆筒状，色泽光亮，薄如纸张，口感鲜美嫩滑，老少皆宜。

Anyue rice roll has a long history and fine workmanship. It is said that at the end of the Qing Dynasty, Meng Ji'an, a native in Anyue, learned the rice roll production technology in Yunnan, and then he came back to Anyue and opened an exclusive shop. Anyue rice roll came out since then. Anyue rice roll selects local high quality water and high quality rice. After soaking the rice, grind it fine with stone mill, hang the pulp, spread into thin pancake, and then steam in the pot. After cooling, wrap it into a roll. All these procedures are done by hand. According to the addition of different raw materials and auxiliary materials, there are white rice roll, yellow rice roll, black rice roll, green rice roll, etc. All kinds of rice rolls are cylindrical, bright in color, as thin as paper, delicious and smooth in taste, and suitable for all ages.

安岳米卷作为特色食材，最常用的烹制方法是凉拌，即将米卷切成段，加入多种调味料拌匀。此外，也采用炒、烧、炸、烤、烫等多种烹饪方法制作菜肴或小吃，风格多样。

As a special food, the most commonly used cooking method of Anyue rice roll is cold salad, which is to cut the rice roll into sections, add a variety of seasonings and mix well. In addition, a variety of cooking methods such as stir-fry, braise, fry, roast, heat are used to make diversified dishes or snacks.

④ 乐至蚕桑 | Lezhi Sericulture

乐至蚕桑，资阳市乐至县著名特产。乐至县是四川主要的蚕桑产业大县、全省第一批优质蚕茧基地县、商务部"东桑西移"项目实施县、国家级蚕桑标准化示范区。

lezhi sericulture is a famous specialty of Lezhi County, Ziyang City. Lezhi County is the main sericulture industry county in Sichuan, the first batch of high-quality silkworm cocoon base county in Sichuan, the implementation county of the "Moving East Mulberry to West" project of the Ministry of Commerce, and the national sericulture standardization demonstration area.

乐至的种桑养蚕历史可追溯到隋唐时期。清光绪三十年（公元1904年），乐至县成立了首个蚕桑业官方机构——蚕桑传习所。1959年，陈毅元帅回家乡视察，曾题诗高度称赞乐至蚕桑的兴盛。1980年代以后，乐至县逐渐发展成为全省乃至全国知名的优质蚕桑基地，经济效益显著。2018年，乐至县桑园面积达10万亩，桑园综合开发9.5万亩，桑园开发和蚕桑副产物综合利用产值达2.85亿元，蚕桑产业的农业部分综合产值达到5.05亿元。

The history of mulberry planting and silkworm rearing can be traced back to the Sui and Tang Dynasties. In the 30th year of Guangxu Period of the Qing Dynasty (1904 A. D.), the first official institution of sericulture industry, Sericulture Teaching and Learning Institute, was established in Lezhi County. In 1959, Marshal Chen Yi returned to his hometown to inspect, and wrote a poem praising the prosperity of sericulture in Lezhi. After the 1980s, Lezhi County gradually developed into a famous sericulture base of high quality in the whole province and even the whole country, with remarkable economic benefits. In 2018, the area of mulberry garden in Lezhi County reached 100,000 mu, the comprehensive development of mulberry garden reached 95,000 mu, the comprehensive utilization output value of mulberry garden development and sericulture by-products reached 285 million yuan, and the comprehensive output value of the agricultural part of sericulture industry reached 505 million yuan.

蚕与桑全身是宝，食用价值高。其中，蚕蛹富含蛋白质和多种氨基酸，是老弱妇孺的高级营养补品，可用炸、炒、炖、卤、煮等烹饪方法制成蚕蛹菜肴。桑叶具有较强的食疗作用，有助于降血压、降血糖等，既用于凉拌、炒、炸等烹饪方法制作桑叶菜点，也制成桑叶茶。此外，用桑叶为原料养殖的桑叶鸡比普通鸡的肉质更细、香味更浓，所产的桑叶蛋比普通鸡蛋的蛋白质含量更高，营养更丰富。桑葚为桑树的成熟果实，甘甜多汁，是常用水果之一，也用于酿制桑葚酒。

The whole bodies of silkworm and mulberry are treasures, with high edible value. Among them, rich in protein and multi-amino acids, silkworm chrysalis is the senior nutritional supplements for the elderly, the weak, women and children, which can be fried, stir-fried, stewed, brined, boiled and cooked through other methods into silkworm chrysalis dishes. Mulberry leaves have a strong therapeutic effect of helping to lower blood pressure and blood sugar, etc., which is not only used for making mulberry leaf dishes with cold salad, stir-frying, frying and other cooking methods, but also for making mulberry leaf tea. In addition, cultivated with mulberry leaves, mulberry leaf chicken has finer meat and better flavor than ordinary chicken, and the mulberry leaf egg produced has higher level of protein and more nutrition than ordinary egg. Mulberry, the ripe fruit of mulberry tree, is sweet and juicy. It is one of the commonly used fruits and also used to make mulberry wine.

⑤ 乐至藕粉 | Lezhi Lotus Root Porridge

乐至藕粉，资阳市乐至县著名特产，是用乐至莲藕加工而成。据《乐至县志》记载，乐至莲藕种植始于宋代，它除了具有肥厚、细嫩、清香的特点外，与其他许多地方的莲藕相比，最大的不同就是多开双花并且具有大小均匀的7孔，故有"花开并蒂，藕贯七心"之说，不产莲子但出粉率高。而其他地方常见的莲藕一般有9～15个孔、多开单花，产莲子但出粉率较低。2018年，乐至县莲藕种植面积达1.8万亩，鲜藕总产量2.98万吨以上，总产值近亿元。近年来，除了乐至，资阳市的雁江区也在大规模种植莲藕，如丹山莲藕也是优质莲藕，属当地有名的食材。

Lezhi lotus root porridge, a famous specialty of Lezhi County, Ziyang City, is processed from Lezhi lotus root. According to the records of *Lezhi Annals*, the planting of Lezhi lotus root began in the Song Dynasty. In addition to the characteristics of being fleshy, delicate and fragrant, compared with lotus root in many other places, the biggest difference of Lezhi lotus root is that it has many double flowers and 7 holes of uniform size. Therefore, there is a saying of "tow flowers growing from the same base, seven hearts running through the lotus root". Lezhi lotus root does not produce lotus seeds but the starch yield is high. The common lotus root in other places generally has 9~15 holes and single flowers, produces lotus seeds but the starch yield is low. In 2018, the planting area of lotus root in Lezhi County reached 18,000 mu, the total output of fresh lotus root was more than 29,800 tons, and the total output value was nearly 100 million yuan. In recent years, besides Lezhi, Yanjiang District of Ziyang City has also been planting lotus root on a large scale. For example, Danzhou lotus root is also of high quality, and is a famous local food material.

乐至县以藕制粉的历史悠久，相传北宋名相寇准在乐至任县令时，曾将香脆可口的莲藕制成色泽鲜亮、味道清香、生津补血的藕粉上贡朝廷。清代乾隆年间，乐至藕粉为皇室贡品。如今，乐至藕粉有两大著名品牌：一是天池藕粉，是中国国家地理标志产品和四川省名牌产品；二是仙荷藕粉，其制作技艺被资阳市人民政府列入市级非物质文化遗产代表性名录，其产品通过ISO 9001质量管理体系认证、HACCP体系认证和国际犹太洁食认证。乐至藕粉色泽白中微红，细腻滑润，常常用沸水冲调后呈半透明胶糊体，芳香甜醇，具有清心明目、养血护肝、美容养颜等作用，尤其适用于老幼人群。

Lezhi County has a long history of making lotus root porridge. According to legend, when the famous Prime Minister of the Northern Song Dynasty Kouzhun was appointed Magistrate of Lezhi County, he made crisp and delicious lotus root into lotus root porridge with bright color and delicate fragrance for royal tribute, which could engender liquid and enrich the blood. During Qianlong Period of the Qing Dynasty, Lezhi lotus root porridge became the royal tribute. Currently, Lezhi lotus root porridge has two famous brands. One is Tianchi Lotus Root Porridge, which is China's National Geographic Indication Product and Sichuan famous brand product. The other is Xianhe Lotus Root Porridge, the manufacturing skills of which have been included in the representative list of municipal intangible cultural heritage by Ziyang Municiple People's Government; its products have passed ISO 9001 Quality Management System Certification, HACCP System Certification and International Kosher Food Certification. Lezhi lotus root porridge is white and slightly red in color, exquisite and smooth. After brewing with boiling water, it is often in semi-transparent paste, which tastes fragrant, sweet and mello. It has the functions of clearing away heart-fire, brightening eyes, nourishing blood, protecting liver and maintaining beauty, which is especially suitable for the old and the young.

⑥ 安岳柠檬 | Anyue Lemon

安岳柠檬，资阳市安岳县特产，中国国家地理标志产品，获得国家绿色食品认证、原产地域产品保护认证，荣获泰国国际博览会金奖、全国名特优农产品评比金奖等多项大奖。

Anyue lemon, a special product of Anyue County, Ziyang City, is China's National Geographic Indication Product, has obtained the national green food certification and protection certification of original region product, and won the gold medal of Thailand International Expo, the gold medal of National Famous, Special, and Excellent Agricultural Products Evaluation and other awards.

安岳被誉为"中国柠檬之都"。全县现有柠檬基地乡镇41个，种植柠檬30余万亩、2 000多万株，约占全国总产量的80%。2018年，安岳柠檬鲜果产量实现58万吨，产值首次突破100亿元，"安岳柠檬"品牌价值达173.6亿元，进入初级农产品类地理标志产品全国10强。

Anyue is known as "Hometown of Lemon in China". The county has 41 lemon base towns, planting more than 300,000 mu of lemon and more than 20 million plants, accounting for about 80% of the total national output. In 2018, the output of fresh fruits of Anyue lemon reached 580,000 tons, and the output value exceeded 10 billion yuan for the first time. The brand value of "Anyue Lemon" reached 17.36 billion yuan, ranking top 10 in China for primary agricultural products of geographical indication products.

安岳柠檬富含柠檬酸和多种维生素、微量元素，具有极强的美容保健作用。安岳人以柠檬鲜果为食材，不仅研发了柠檬系列菜品，如柠檬鸡豆花、柠香排骨、柠檬烤鱼、柠檬酥排骨等，还以柠檬为主题研发了安岳柠檬宴，在2018年9月被评为"中国菜"之四川主题名宴。除了直接使用、销售鲜果外，安岳还对柠檬进行精深加工。安岳县的柠檬产业发达，现有柠檬加工企业27家，其中国家级和省级重点龙头企业各1家、市级重点龙头企业5家，年加工能力30万吨，生产开发出柠檬油、柠檬酒、柠檬醋、柠檬茶和柠檬饮料等柠檬制品18大类、39个品种。柠檬鲜果及其加工产品远销美国、法国、波兰、德国、加拿大、俄罗斯、哈萨克斯坦等30多个国家和地区。

Anyue lemon is rich in citric acid and a variety of vitamins, microelements, which has a strong function of beauty maintaining and health care. Taking lemon fresh fruit as food material, Anyue people not only created a series of lemon dishes, such as Lemon Flavored Chicken Doufu, Lemon-Flavored Fried Spareribs, Lemon Grilled Fish, and Lemon Crispy Ribs and so on, but also developed Anyue Lemon Banquet with lemon as the theme, which was named as Sichuan Famous Theme Banquet of "Chinese Cuisine" in September, 2018. In addition to directly using and selling fresh fruits, Anyue also carries on intensive processing of lemon. The lemon industry in Anyue County is flourishing. There are 27 lemon processing enterprises in Anyue County, including 1 national and 1 provincial key leading enterprises and 5 municipal key leading enterprises, with an annual processing capacity of 300,000 tons. The county has developed 18 categories and 39 varieties of lemon products such as lemon oil, lemon wine, lemon vinegar, lemon tea and lemon drinks. Lemon fresh fruit and its processed products are exported to more than 30 countries and regions such as the United States, France, Poland, Germany, Canada, Russia, Kazakhstan, etc.

⑦ 安岳通贤柚 | Anyue Tongxian Pomelo

通贤柚，资阳市安岳县特产，中国国家地理标志产品，1992年12月被评为四川省"良种柚第一名"和"全国优质柚类第三名"，曾获得全国农业博览会金奖等多项奖励。

Tongxian pomelo is a special product of Anyue County, Ziyang City, and China's National Geographic Indication Product. In December, 1992, Tongxian pomelo was awarded "The First Prize of Well-Bred Pomelo" and "The Third Prize of National High Quality Pomelo" of Sichuan province. And it has once won the gold medal of National Agricultural Exposition and other awards.

通贤柚于20世纪30年代从福建漳州地区引种到安岳县通贤镇。经过不断发展，尤其是近10余年来的大力发展，安岳县的通贤柚集中在6个乡镇上，并且建立了通贤柚科技示范园，现已形成了800万株、10万亩、5万吨的生产规模和50公里长廊，预计盛产期产量可达30万吨。通贤柚销售网络健全，县内建有专业批发市场2个，并在59个城市设立了销售网点，产品远销国内136个大中城市。通贤柚现已形成产、供、销一条龙的产业化模式，成为产区农民的"致富果"。

Tongxian pomelo was introduced to Tongxian Town of Anyue County from Zhangzhou area of Fujian Province in the 1930s. After continuous development, especially the vigorous development in the past 10 years, Anyue Tongxian pomelo has covered 6 towns and villages, and Tongxian Pomelo Science and Technology Demonstration Park has been established, forming the 8 million plants, 100,000 mu and 50,000 tons of production scale and 50 kilometers of corridor, with an expected peak production of 300,000 tons. The sales network of Tongxian pomelo is sound. There are two professional wholesale markets in the county. The sales networks have been set up in 59 cities. The products are exported to 136 large and medium-sized cities in China. Tongxian pomelo has formed an industrialization mode of production, supply and marketing, and has become the "rich fruit" of the farmers in the production areas.

通贤柚果实呈倒卵形，单果重1 000～1 500克，果色橙黄，果皮薄，果肉晶莹剔透、无核，香味浓郁、甜酸适度、汁多脆嫩，富含维生素C及钙、铁、锌、镁等矿物质，具有润肺化痰、生津止渴、增进食欲、提神醒脑等作用，为果中珍品。

The fruit of Tongxian pomelo is obovate, the weight of a single fruit is 1,000~1,500 grams the fruit color is orange, the skin is thin, the flesh is crystal clear, seedless, and the flavor is strong, with moderate sweet and sour, crisp, tender and juicy. Being praised as the treasure in fruits, it is rich in vitamin C and minerals such as calcium, iron, zinc, magnesium, etc. with the functions of moistening lung for removing phlegm, engendering liquid and allaying thirst, improving appetite, refreshing, etc.

⑧ 雁江蜜柑 | Yanjiang Sweet Mandarin

雁江蜜柑，资阳市雁江区特产，中国国家地理标志产品，曾获得四川省优质水果称号。2005年，雁江蜜柑种植园被批准为"国家级农业标准化示范区""省级优势产业核心示范区"和"国家级无公害标准化生产示范区"。

Yanjiang sweet mandarin, a special product in Yanjiang District of Ziyang City and China's National Geographic Indication Product, has been awarded the title of Sichuan High-quality Fruit. In 2005, Yanjiang Sweet Mandarin Plantation was approved as "National Agricultural Standardization Demonstration Area", "Provincial Core Demonstration Area of Advantageous Industries" and "National Pollution-Free Standardized Production Demonstration Area".

雁江区为长江中上游地区规模最大、全省唯一的早熟蜜柑产销基地，柑橘种植面积达19.5万亩，产量17.1万吨，全省排名第4，品牌价值8.53亿元。雁江蜜柑按成熟期来分有两类：一是早熟蜜柑，主要品种有兴津蜜柑、宫本等；二是晚熟杂柑，主要有大雅、不知火、春见等品种。由于特定的土壤、水质、气候等因素，通过科学栽培，雁江蜜柑优良且独特的品质，其果实扁圆、形状端正，油胞细密，光滑而富有弹性，成熟早、着色早、酸甜适度、果汁丰富，还具有促进食欲、醒酒等多种作用。

Yanjiang District is the largest in the middle and upper reaches of the Yangtze River and the only in Sichuan Province of the early-maturing sweet mandarin production and marketing base, with a planting area of 195,000 mu and an output of 171,000 tons, ranking the 4th in the province with a brand value of 853 million yuan. There are two kinds of Yanjiang sweet mandarins according to the mature stage. One is the early-maturing sweet mandarin, the main varieties are Xingjin sweet mandarin, Miyamoto sweet mandarin and so on; the other is the late-maturing mandarin hybrids, mainly include Daya, Buzhihuo, Chunjian and other varieties. Due to the specific soil, water quality, climate and other factors, through scientific cultivation, Yanjiang sweet mandarin has excellent and unique quality. The fruit is oblate with regular shape and fine oil cells, smooth and elastic, maturing early, coloring early, moderate sweet and sour with rich fruit juice. It also has a variety of effects, such as improving appetite, sobering up, etc.

2 LIVESTOCK AND AQUATIC PRODUCTS 禽畜水产类

① 乐至黑山羊 | Lezhi Black Goat

乐至黑山羊，资阳市乐至县特产，全国农产品地理标志。2010年农业部将川中黑山羊（乐至型）纳入畜禽遗传资源目录。乐至县是四川省养羊十强县、首批无公害肉羊生产基地县、优质肉羊生产基地县、国家级秸秆氨化养羊示范县、国家级农业标准化（黑山羊）示范区、国家无规定疫病区示范县，被誉为"中国黑山羊之乡"。

Lezhi black goat, a specialty in Lezhi County of Ziyang City, is China's National Geographic Indication Product. In 2010, the Ministry of Agriculture included mid-Sichuan black goat (Lezhi variety) in the Catalogue of Livestock and Poultry Genetic Resources. Lezhi County is one of the Top Ten Sheep Breeding Counties in Sichuan, one of the first batch of Pollution-Free Meat Sheep Production Base Counties, one of the High-Quality Meat Sheep Production Base Counties, one of the National Demonstration Counties of Breeding Sheep with Ammoniated Straw, one of the National Agricultural Standardization (Black Goat) Demonstration Areas, and one of the National Demonstration Counties of Disease Free Zone. It is honored as the "Hometown of Black Goat in China".

乐至黑山羊的养殖历史悠久。清代道光年间的《乐至县志》载："惟黑山羊，纯黑味美，不膻。"经过长期的自然选择和人工培育，乐至黑山羊形成了适应性强、前期生长发育快、产肉性能好、繁殖性能突出、遗传性稳定等优良特性，全身毛被黑色、富有油亮光泽，体型大，背腰平直，肉色红润、肉质细嫩，香而不腻，具有山羊肉特有的香味。截至2018年底，全县存栏黑山羊33.3万只，出栏山羊59.16万只，活羊及羊产品远销全国20余个省、市，产值11.53亿元，占全县畜牧业产值30%以上。

Lezhi black goat has a long history of breeding. During Daoguang Period of the Qing Dynasty, *Lezhi Annals* recorded that "only the black goats are pure black and taste good, with no smelly odor of mutton." After a long period of natural selection and artificial cultivation, Lezhi black goat has formed the prominent characteristics of well adaptability, rapid early growth, good meat production and outstanding reproductive performance, stable hereditary stability, etc. Being covered by the black, bright and shiny hair on the whole body, Lezhi black goat has big body, straight back and waist, red and tender meat which tastes fragrant but not greasy with the unique flavor of goat meat. By the end of 2018, the amount of breeding black goats in the county was 333,000 and the number of black goats sold to the market was 591,600. Live goats and goat products were sold to more than 20 provinces and cities in China. The output value of goats was 1.153 billion yuan, accounting for more than 30% of the county's animal husbandry output value.

近年来，乐至黑山羊产业链得到不断拓展和完善。从事羊肉经营的餐馆、酒店蓬勃发展，羊肉干、羊肉香辣酱、速冻羊肉汤等产品相继问世。此外，乐至已成功举办九届"乐至烤肉美食节"，烤羊肉串、烤羊排、烤全羊等美食得到消费者的青睐。

In recent years, Lezhi black goat industry chain has been continuously expanded and improved. Restaurants and hotels engaged in mutton business are booming, and products such as mutton jerky, mutton hot spicy sauce, quick-frozen mutton soup have come out one after another. In addition, Lezhi has successfully held nine "Lezhi Pork BBQ Food Festivals". Mutton shashlik, grilled lamb chops, roasted whole lamb and other delicacies have been favored by consumers.

② 伍隍猪 | Wuhuang Pig

伍隍猪，资阳市雁江区伍隍镇特产，是优良地方猪种之一和重要的"资味"农业区域品牌，已纳入国家畜禽遗传资源目录，国家工商总局也已受理"伍隍猪"三个字注册。

Wuhuang pig, a special product of Wuhuang Town, Yanjiang District, Ziyang City, is one of the excellent local pig species and an important agricultural regional brand of "Zi Wei", which has been included in the Catalogue of Livestock and Poultry Genetic Resources. The State Administration of Industry and Commerce (SAIC) has also accepted the registration of "Wuhuang Pig"

伍隍猪适应性、抗逆性和抗病能力强，耐粗饲，生长速度快；母性好，产仔率高，平均每窝产仔14头；瘦肉率达50%，是我国地方品种中瘦肉率最高的品种，肉质口感好。伍隍镇党委、镇政府高度重视伍隍猪选种保育工作，采取措施推进其产业差异化、特色化、品牌化发展，积极促进民众养殖伍隍猪，提升其养殖规模。伍隍猪是制作糖醋里脊、红烧肉、水煮肉片、炸酥肉等菜品不可多得的理想食材，凭借其优势成为了大众餐桌上非常受欢迎的美食。

Wuhuang pig has strong adaptability, stress resistance and disease resistance. It is resistant to coarse feeding and grows fast. It has good maternity, high farrowing rate, and an average of litter size is 14. The lean meat rate reaches 50%, which is the highest among local varieties in China. And the meat tastes good. The party committee and government of Wuhuang Town attach great importance to the strain selection and conservation of Wuhuang pig, take measures to promote the development of industrial differentiation, specialization and branding, actively accelerate the public breeding of Wuhuang pig, and expand the breeding scale. Wuhuang pig is a rare and ideal food material for making Fried Sweet and Sour Tenderloin, Stewed Pork with Brown Sauce, Boiled Pig Meat Slice, Crispy Deep Fried Pork and other dishes. With the advantages of Wuhuang pig, these dishes have become the very popular foods on the table.

③ 坛子肉 | Tanzi Pork

坛子肉，资阳市的著名特产，在资阳市的各个县、区都有制作。其中，乐至外婆坛子肉制作技艺被资阳市人民政府列入市级非物质文化遗产代表性名录；安岳普州坛子肉被列为"四川省地方名优产品推荐名目"，被资阳市评为"资阳名料"，并成功申报国家发明专利。

Tanzi Pork, a famous specialty of Ziyang City, is made in the counties and districts of Ziyang City. Among them, the making techniques of Lezhi Diced Pork in Grandma's Pot were included on the representative list of municipal intangible cultural heritage by Ziyang Municipal People's Government; Puzhou Tanzi Pork of Anyue was listed on "Excellent Local Products Recommended Items in Sichuan", and was rated as "Ziyang Famous Material" by Ziyang City and successfully got a national invention patent.

在以往食物短缺和储藏条件受到限制的年代，资阳先辈将豇豆干、青菜干等各种干菜和猪肉拌上五香、八角等香料，以一层干菜、一层猪肉的顺序铺入土坛中腌渍数月，以达到储存、防腐和便于食用的目的。这种用坛子腌渍的肉，俗称坛子肉，是经乳酸菌厌氧发酵而成，辅以盐菜的清香，不同于酱肉制品和烟熏腊肉，更加原汁原味、醇香可口、方便食用。制作坛子肉，需要5道工序：一是选肉，一定要选带有肥肉的肉，由此做出来的坛子肉才香；二是切肉，根据使用量和习惯，将肉切成长方块、洗净并晾干；三是煎肉，将晾干的肉块放入热锅里煎；四是码盐，将肉均匀涂抹上盐后放置两三天；五是入坛，将码好盐的肉和干菜，按照一层肉、一层干菜的顺序放入坛子中密封好即可。

In the past, when food was in shortage and the storage conditions were limited, Ziyang ancestors mixed dried cowpea, dried green vegetables and pork with spices such as five spices and aniseed, and put them into the earthen pot in the order of one layer of dried vegetables and one layer of pork for several months to pickle, so as to achieve the purpose of storage, preservation and being convenient to eat. This kind of pickled meat is commonly known as Tanzi Pork. It is produced by lactic acid bacteria anaerobic fermentation, and is supplemented by the flavor of salty vegetables. Different from sauced meat products and preserved meat, it is more original, mellow, delicious, and convenient to eat. There are five steps to make Tanzi Pork. The first is to choose the meat. Make sure to choose the meat with fat on it, and then the Tanzi Pork will be fragrant. The second is to cut the meat. According to the usage amount and habits, cut the meat into cubes, wash and dry. The third is to fry the meat. Fry the dried meat cubes in a hot pan. The fourth is to salt the meat. Salt the meat evenly and leave it for two or three days. The fifth is to store in the pot. Put the salt meat and dried vegetables into the pot in the order of one layer of meat and one layer of dried vegetables and seal.

坛子肉营养丰富、风味独具一格，作为特色食材在菜点制作中得到广泛运用，常采用蒸、炒、焖、炖等烹饪方法，其常见菜点有旱蒸坛子肉、坛子肉粑粑汤等。

Tanzi Pork is rich in nutrition and unique in flavor. As a characteristic food, it is widely used in the production of dishes. Cooking methods such as steaming, stir-frying, braising and stewing are often adopted. Common dishes include Steamed Tanzi Pork and Soup of Baba Cakes with Tanzi Pork, etc.

④ 龙洞湾乳鸽 | Longdongwan Squab

龙洞湾乳鸽，资阳市雁江区特产。雁江区生态环境优美，吸引了每年南北迁徙的大雁在此中途休息补养，故得名雁江，又名雁城。龙洞湾乳鸽即生长于如此优越的生态环境之中，其基地建于山顶，通风良好，阳光充足，清洁卫生，其饲料为优质玉米、高粱、豌豆、小麦等，保证了乳鸽肉的品质。近年来，龙洞湾乳鸽养殖产业加快发展，计划年出栏100多万对，年产值可达3千多万元，正努力打造四川最大的乳鸽养殖产业园。

Longdongwan Squab, a specialty in Yanjiang District of Ziyang City. Yanjiang District has a beautiful ecological environment, attracting the wild geese to rest and recuperate here in the midway of north-south migration; hence the district is named Yanjiang and also known as Yancheng. Longdongwan squab is growing in such a superior ecological environment. Its base is built on the top of the mountain, with good ventilation, sufficient sunshine, clean surroundings. Its fodders include high quality corn, sorghum, pea, wheat, etc. to ensure the quality of pigeon meat. In recent years, the development of pigeon breeding industry in Longdongwan has been accelerated. It plans to produce more than 1 million pairs of squab every year, with an annual output value of more than 30 million yuan. It is striving to build the largest squab breeding industrial park in Sichuan.

龙洞湾乳鸽肉质细嫩，营养丰富，经测定，含有17种以上氨基酸、总量达53.9%，且含有10余种微量元素及多种维生素，是高蛋白、低脂肪的理想食品，具有养血补气、消除疲劳等作用。乳鸽的烹饪方式多样，常见菜品有炸乳鸽、油焖乳鸽、清炖乳鸽、泡椒乳鸽等。

Longdongwan squab has tender meat and various nutriments. Via determination, it contains more than 17 kinds of amino acids and the total amount reaches 53.9%. It also contains more than 10 kinds of microelements and various vitamins. Thus it is an ideal food with high protein and low fat, which can nourish blood and replenish qi and eliminate fatigue. Pigeon can be cooked in a variety of ways. Common dishes include Fried Squab, Braised Squab, Stewed Squab in Clear Soup, and Squab with Pickled Peppers, etc.

⑤ 乐至白乌鱼 | Lezhi White Snakehead

乐至白乌鱼，资阳市乐至县特产，中国国家地理标志产品。白乌鱼，又名甲乌鳢，俗称白乌棒，隶属鲈形目鳢科鳢属淡水名贵经济鱼类，是世界上的一种稀有水产品种，有鱼中珍品之称。在乐至，白乌鱼的养殖遍布25个乡镇，繁养基地达6 000余亩，年产白乌鱼苗2 000万尾、成鱼2万吨，年产值近10亿元。

Lezhi white snakehead, a special product of Lezhi County, Ziyang City, is China's National Geographic Indication Product. White snakehead belongs to ChaIlnidae of Perciformes and is the freshwater rare economic fish. It is a rare aquatic species in the world, which is known as the treasure in the fish. In Lezhi, white snakehead is cultivated in 25 villages and towns, with a breeding base of more than 6,000 mu. The annual output of white snakehead fry is 20 million, of adult fish is 20,000 tons and the annual output value is nearly 1 billion yuan.

乐至白乌鱼体两侧的鳞片呈色白，背部呈灰白色或点状褐色，在水体中呈银白色，头部扁平，头部鳞片较大，尾部鳞片细，鳍条呈金黄色，在水中游姿优美，动作简单、快捷，可作观赏鱼，其肌肉紧密、有弹性、白色，无肌肉间刺。乐至白乌鱼肉质细嫩、味道鲜美，除含DHA、无机盐、维生素外，还含有17种氨基酸，其中包括8种人体必需氨基酸，具有氨基酸和蛋白质含量高、脂肪含量低的特点，营养价值极高，具有滋补调养、促进人体生肌补血、愈合伤口等独特的作用。

The scales on both sides of Lezhi white snakehead's body are white, the back is grayish white or point-like brown, and it is silver-white in water. The head is flat, the scales on the head are large and those on the tail are thin, and the fins are golden. It is graceful when swimming, simple and quick in action, and can be ornamental fish. Its muscles are tight and elastic, white, without intermuscular spines. Its meat is tender and delicious. Besides DHA, inorganic salt, vitamins, it also contains 17 kinds of amino acids, including 8 kinds of essential amino acids for human body. And it has the characteristics of containing high amino acids and protein and low fat. In addition, it is of high nutritional value and has the distinctive effects of nourishing and nursing, promoting the body to build the muscles and enrich the blood, and helping heal wounds and so on.

乐至白乌鱼是一种较为珍贵的特产食材，常采用炒、熘、焖、炖等烹饪方法制作菜肴，其常见品种有生炒鱼片、白菊鱼丝、白乌鱼炖土鸡等。

Lezhi white snakehead is a kind of precious specialty food, which is often cooked by frying, quick-frying, braising, stewing and other methods. The common dishes include Fried Fish Slices, Shredded Fish with White Chrysanthemum, White Snakehead and Chicken Stew, etc.

⑥ 雁江泥鳅 | Yanjiang Loach

雁江泥鳅，资阳市雁江区特产。自古有"天上斑鸠，地下泥鳅"的说法，泥鳅的蛋白质含量在肉类食品中名列前茅，是高蛋白、低脂肪、极低胆固醇的佳品，有美容养颜、改善贫血、软化血管、增强免疫力、强健筋骨等作用。

Yanjiang loach is a specialty in Yanjiang District, Ziyang City. Since ancient times, there exists the reputation of "turtledove in the sky, loach on the ground". The protein content of loach is among the highest of meat, and it is a wonderful product with high protein, low fat and extreme low cholesterol. It can maintain beauty, improve anemia, soften blood vessels, enhance immunity and strengthen muscles and bones, etc.

雁江水域面积30余万平方公里，水源充足，沟壑纵横，土地肥沃，光照充足，周边无工业污染，有得天独厚的泥鳅养殖条件，所产泥鳅成为资阳主要的经济鱼类，其养殖技术及规模处于全省前列。马蹄湾水产养殖专业合作社主要从事泥鳅家系选种、育种、规模化繁育工作，是农业部健康养殖示范场和四川省无公害水产品产地，面向国内年生产、销售泥鳅优良苗种约100亿尾，生产销售泥鳅700吨。资阳民间长期食用泥鳅，制作的泥鳅菜具有糯、酥、香、鲜、嫩等特点。

The water area of the Yanjiang River is more than 300,000 square kilometers, with abundant water, crisscrossed ravines and gullies, fertile land and sufficient sunlight. There is no industrial pollution in the surrounding areas, and it has unique breeding conditions for loach. The loach produced by Yanjiang River has become the main economic fish in Ziyang, and its breeding technology and scale are in the forefront of the whole province. Matiwan Specialized Aquaculture Cooperatives is mainly engaged in species selection, breeding and large-scale breeding of loach families. It is a Healthy Breeding Demonstration Farm of the Ministry of Agriculture and a source of pollution-free aquatic products in Sichuan. It produces and sells about 10 billion fine loach seedlings and 700 tons of loach annually in China. Ziyang has a long history of loach eating, and the loach dishes have the features of soft, crisp, tasty, fresh, tender, etc.

3 调味品及速冻食品类 | SEASONINGS AND QUICK-FROZEN FOODS

① 临江寺豆瓣 | Linjiangsi chili bean paste

临江寺豆瓣，资阳市雁江区著名特产，中国国家地理标志产品。临江寺豆瓣制作技艺被四川省人民政府列入省级非物质文化遗产代表性名录，"临江寺"系列产品还先后获得"中华老字号""中国驰名商标""中国名优产品""四川名牌""四川省著名商标"等称号。

Linjiangsi chili bean paste, a famous specialty in Yanjiang District of Ziyang City, is China's National Geographic Indication Product. The production techniques of Linjiangsi chili bean paste have been included in the representative list of provincial intangible cultural heritage by the People's Government of Sichuan Province. Linjiangsi Temple series products have also won in succession the titles of "China Time-Honored Brand", "China Renowned Brand", "China Famous and Excellent Product", "Sichuan Famous Brand", "Sichuan Famous Trademark" and so on.

临江寺豆瓣传统工艺始创于清乾隆三年（公元1738年），历经八代传承。临江寺豆瓣选用当地的良种蚕豆和芝麻为主料，配以食盐、花椒、胡椒、白糖、金钩、火肘、鸡松、鱼松、香油、红曲、辣酱、麻酱、甜酱，以及多种香料精工酿制而成。加工时要经过蚕豆脱壳、浸泡、接种、制曲、撒盐水等多道工序，再进行发酵，最后与各种辅料按比例配制，便成为成品豆瓣酱。临江寺豆瓣，色泽红润，瓣粒成型而柔和化渣，味道微辣回甜、油而不腻、辣而不辛、咸而不涩、酱香浓郁、回味悠长，营养丰富。临江寺豆瓣历经数百年的市场考验，已成为成渝古道上驰名川内外的地方特产，是资阳人引以自豪的具有丰富历史文化内涵的知名品牌。

The traditional craft of Linjiangsi chili bean paste was created in the 3rd year of Qianlong Period (1738 A. D.) of the Qing Dynasty and has been passed on for eight generations. Linjiangsi chili bean paste uses local fine breed broad beans and sesame seeds as main materials, with salt, Sichuan pepper, pepper, sugar, dried shrimp, ham, chicken

floss, fish floss, sesame oil, red yeast rice, chili sauce, sesame paste, sweet sauce and various spices as the ingredients to brew precisely. During processing, it has to go through many procedures, such as broad bean hulling, soaking, inoculation, starter-making, salt water spreading, etc., and then ferment, and finally prepare in proportion with all kinds of auxiliary materials to become the finished chili bean paste. Linjiangsi chili bean paste is ruddy in color, and its flaps are shaped and soft. The taste is slightly spicy with a back to the sweet, oily but not greasy, spicy but not pungent, salty but not bitter, with a strong flavor and a long aftertaste and rich nutrition. After hundreds of years of market test, it has become a well-known local specialty inside and outside of Sichuan. It is a famous brand with rich historical and cultural connotation that Ziyang people are proud of.

临江寺豆瓣系列产品有香油豆瓣、金钩豆瓣、火肘豆瓣、红油豆瓣、鱼松豆瓣等50多个品种、180多种规格，畅销20多个省、市、自治区，远销美国、日本、加拿大等国家，是川菜上等的调味品，广泛用于炒菜、烧菜、水煮系列菜肴及火锅调味。

Linjiangsi chili bean paste series products consist of more than 50 varieties and over 180 specifications, such as sesame oil bean paste, dried shrimp bean paste, ham bean paste, chili oil bean paste, fish floss bean paste and so on, which sell well in more than 20 provinces, and are exported to the United States, Japan, Canada and other countries and regions. Chili bean paste is the best seasoning in Sichuan cuisine, and is widely used for stir-frying, braising, and boiling series of dishes, and hot pot seasoning.

② 中和醋 | Zhonghe Vinegar

中和醋，资阳市雁江区著名特产，具有近半个世纪的历史，被认定为蜀中佳醋、传统名特食品，入选首届四川食品博览会消费者最喜爱产品。

Zhonghe vinegar, a famous specialty in Yanjiang District, has a history of nearly half a century. It is recognized as the Outstanding Vinegar in Sichuan and the Traditional Famous Special Food, and is selected as The Most Favorite Product of Consumers in The First Sichuan Food Expo.

资阳雁江区的中和镇素以"酿造之乡"著称,其出产的中和醋、中和酱油、中和榨菜等远近驰名。其中,中和醋以纯粮酿造、天然晒露、利用微生物科学发酵而成,味道酸中带有回甜,醇厚绵长,醋香浓郁,含有10余种人体必需的氨基酸、多种维生素和锌、铜、铁、磷、钾等10种微量元素,具有增进食欲、平血压等作用,远销北京、上海、重庆、成都等省内外大中城市。

Zhonghe Town in Yanjiang District of Ziyang City is known as the "Hometown of Brewing", and its Zhonghe vinegar, Zhonghe soy sauce, Zhonghe preserved mustard, etc. are well-known far and near. Among them, Zhonghe vinegar is brewed by pure grains, with natural sun exposure, fermented by microorganism science. It tastes sour with a back to the sweet, mellow and lingering, and has strong vinegar flavor. As containing more than ten kinds of amino acids necessary for human body, a variety of vitamins and 10 kinds of microelements such as zinc, copper, iron, phosphorus, potassium, etc. Zhonghe vinegar has the effects of improving appetite, smoothing blood pressure and so on. It is exported to Beijing, Shanghai, Chongqing, Chengdu and other large and medium-sized cities in and out of the province.

中和醋是制作川味菜肴、面点小吃常用的酸味调料,深受广大川菜经营企业和消费者喜爱。

Zhonghe vinegar is a sour condiment commonly used in Sichuan dishes and snacks, and is deeply popular with Sichuan cuisine business enterprises and consumers.

③ 安岳紫竹姜 | Anyue Zizhu Ginger

紫竹姜，资阳市安岳县著名特产，又名鼎新姜，因产于安岳县紫竹观音附近的鼎新乡而得名。2003年2月，紫竹姜及其生产基地被四川省农业厅列为"无公害农产品"和"无公害农产品生产基地"。

Zizhu ginger, a famous specialty in Anyue County of Ziyang City, also known as Dingxin ginger, gets the name because it is produced in Dingxin Township, near Zizhu Guanyin of Anyue County. In February, 2003, Zizhu ginger and its production base were listed as "Pollution-Free Agricultural Product" and "Production Base for Pollution-Free Agricultural Products" by Sichuan Provincial Department of Agriculture.

在安岳，生姜的种植历史可上朔到明清时期。如今，安岳的生姜产业化已初具规模，主要以鼎新乡、天林镇为主，共计12个乡镇成片种植，全县种植面积万余亩，四季都有出产，年产量约3万吨，产值近3亿元。安岳是成渝等地生姜供应的主要产地，也长期被内江三元泡菜厂确定为生姜首选生产基地，其生姜远销重庆、新疆、宁夏、贵州、云南等省区。

In Anyue, ginger planting history can date back to the Ming and Qing Dynasties. Nowadays, the ginger industrialization of Anyue has begun to take shape, mainly in Dingxin Township and Tianlin Town. A total of 12 townships are known for patch planting, covering a planting area of more than 10,000 mu in the county. There are outputs in all seasons, with an annual output of about 30,000 tons and an output value of nearly 300 million yuan. Anyue is the main source of ginger supply in Chengdu, Chongqing and other places, and has long been identified as the preferred production base for ginger by Neijiang Sanyuan Pickle Factory. The ginger is exported to Chongqing, Xinxiang, Ningxia, Guizhou, Yunnan and other provinces and regions.

紫竹姜具有芽长筋少、质地脆嫩、味美清香等特点，品质优良，既是人们喜爱的蔬菜，也是烹饪中不可缺少的调味料，特别适用于制作隔夜泡姜和仔姜系列菜肴，还具有一定的防病治病作用。

Zizhu ginger has the characteristics of long bud and little tendon, crisp texture, good flavor and fragrance, and good quality, etc. It is not only a favorite vegetable for people, but also an indispensable seasoning in cooking, and it is especially suitable for making overnight pickled ginger and tender ginger series of dishes. It also has a certain effect of disease prevention and treatment.

④ 乐至青花椒 | Lezhi Green Sichuan Pepper

乐至青花椒，资阳市乐至县著名特产。其中，帅青花椒是乐至青花椒的著名品牌，荣获"资阳市知名商标""四川省名牌产品"等称号。

Lezhi green Sichuan pepper is a famous specialty of Lezhi County, Yiyang City. Shuai Green Sichuan Pepper is a famous brand of Lezhi green Sichuan pepper, which has won the titles of "Ziyang Well-known Trademark" and "Sichuan Famous Brand Product".

乐至青花椒色泽翠绿，口味清爽，香味浓郁、独特而持久，果粒圆大，椒口像鱼嘴、多呈双耳，果皮肉厚、有光泽，表面油囊密生、鼓实，凸起的精油腔多而密，内果皮光滑、多数与果肉分离而卷曲，果实上多为并蒂附生1~3粒纯肉小椒粒。含油多、香气浓、麻味足，是川菜制作中重要的麻味调料和香料，有去腥增味的作用。

Lezhi green Sichuan pepper has green color, fresh taste, and rich, unique and lasting fragrance. The fruits are round and big, the pepper mouth is like a fish mouth, commonly with two ears, the skin is thick and glossy. The oil sacs on the surface are densely packed and plump, and the raised essential oil cavities are numerous and dense. The endocarp is smooth, most of them separate from flesh and curl. Most of the fruits are double epiphytic 1~3 small pepper corns with pure meat. It is rich in oil, strong in aroma and sufficient in numbing taste. It is an important numbing flavor seasoning and spice in the production of Sichuan cuisine, and has the function of deodorizing and flavoring.

乐至青花椒不仅直接用于川菜制作，也进行精深加工。其中，帅青花椒开发有限公司作为省级林业产业化龙头企业，每年加工贮存鲜青花椒近5 000吨，年产花椒油等系列调味油1 000余吨，实现产值5 500万元。此外，资阳市境内其他地方也在种植和生产优质青花椒，如安岳县裂嘴麻青花椒也较为知名。

Lezhi green Sichuan pepper is not only directly used in making Sichuan cuisine, but also in deep processing. As the provincial leading enterprise of forestry industrialization, Shuai Green Sichuan Pepper Development Co., Ltd. processes and stores nearly 5,000 tons of fresh green Sichuan pepper every year, and produces more than 1,000 tons of pepper oil and other series of seasoning oil annually, achieving an output value of 55 million yuan. In addition, there are other places growing and producing high-quality green Sichuan pepper in Ziyang City, for example, Liezuima Green Sichuan Pepper of Anyue County is also comparatively famous.

⑤ 安井速冻食品 | Anjing Quick-Frozen Food

速冻食品，通常是指通过急速低温（-18℃以下）冻结并包装、储存、送抵消费地点的低温食品，种类较多，有水产速冻食品、农产速冻食品、畜产速冻食品和调理类速冻食品。其中，调理类速冻食品，特指两种以上的原料经加工处理后急速低温冻结而成的食品，又包括中式点心类、火锅调料类、菜肴类等。一般而言，速冻食品的质量大多高于缓冻食品，其最大优点是可以通过低温来保存食品原有品质、最大限度地保存其营养，而无需添加任何防腐剂和添加剂，更加美味、方便、健康、卫生、营养。

Quick-frozen food usually refers to the low-temperature food frozen through rapid low temperature (below -18℃), and then packaged, stored and sent to the places of consumption. There are many kinds of quick-frozen food, such as aquatic quick-frozen food, agricultural quick-frozen food, livestock quick-frozen food and prepared quick-frozen food, etc. Among them, the prepared quick-frozen food refers in particular to the food which is frozen rapidly at low temperature after processing of more than two kinds of raw materials, including Chinese dim sum, hot pot seasonings, dishes and cuisine and so on. Generally speaking, the quality of quick-frozen food is mostly higher than that of slow-frozen food. Its biggest advantage is that it can maximize the preservation of the original quality of food and its nutrition through low temperature, without adding any preservatives and additives. Thus the food is more delicious, convenient, healthy, hygienic and nutritious.

资阳速冻食品的种类较为丰富，常见速冻食品供应充足，是资阳美食重要的特色食材。在资阳众多速冻食品品牌中，四川安井食品有限公司最具代表性。该公司位于资阳市雁江区医药食品产业园，创建于2016年5月，厂区占地面积127.89亩，设计年产15万吨速冻食品的生产规模，其中，肉制品9万吨，米面制品1万吨，鱼糜制品5万吨，正在规划豆制品、调理肉制品（预制菜肴制品）落地四川基地生产。届时，该公司的速冻食品将涵盖传统火锅料制品、米面制品、调理肉制品、豆制品等，可以进一步推动资阳速冻食品的发展。

As an important special ingredient of Ziyang cuisine, there are many kinds of quick-frozen foods in Ziyang, and the supply of common quick-frozen food is sufficient. Among many brands of quick-frozen food in Ziyang, Sichuan Anjing Food Co., Ltd. is the most representative. Founded in May, 2016, the company is located in the Medicine and Food Industrial Park of Yanjiang District in Ziyang City, covering an area of 127.89 mu, designing the production scale of annual output of 150,000 tons of quick-frozen food, among which, 90,000 tons of meat products, 10,000 tons of rice and flour products and 50,000 tons of minced fish products. The quick-frozen food of bean products and prepared meat products (prepared dishes products) are being planned to be produced in Sichuan base. By then, the company's quick-frozen food will cover traditional hot pot ingredients products, rice and flavor products, prepared meat products, bean products, etc., which can further promote the development of the quick-frozen food in Ziyang.

Chapter Three

RENOWNED DISHES AND SNACKS OF ZIYANG

Endowed with a mild climate, Ziyang City boasts evergreen vegetation and abundant produce. Intelligent and diligent local people are good at using native ingredients in their cuisine, and delicious dishes and snacks have been created in its long culinary history. The advance of social development and the improvement of people's living standards have helped to increase its dish varieties, and diners and chefs have begun to pursue foods that are stylish, green, safe, and nutritious. Ziyang cuisine is now an important part of Sichuan dishes and a significant trademark for Ziyang City. Because of the limited length of the book, only one hundred dishes and snacks are chosen here in accordance with the following three criteria: Firstly, the dishes are peculiar to Ziyang in terms of cooking techniques, flavors and history. Secondly, they utilize local ingredients, combining the first, second and tertiary industries and contributing to rural rejuvenation and poverty alleviation efforts. Thirdly, they have high cultural values and strong tourism appeal, integrating tradition with innovation, globalization with localization, and peculiarity with universalism, which are conducive to promoting Ziyang's economic and social growth as well as satisfying people's need for a better life.

第三篇 资阳名特菜肴与面点小吃

资阳市四季常青,物产丰富,一代代勤劳智慧的资阳人将本地食材灵活巧妙地制作成珍馐佳肴,令人齿颊留香,流连忘返。随着社会发展和人民生活水平的提高,资阳美食琳琅满目,数不胜数,并且愈发追求安全、绿色、营养、时尚,成为资阳文化与旅游的靓丽名片,是川菜文化宝库中的一颗璀璨明珠。但是,由于篇幅所限,这里仅选取其中的名特菜肴与面点小吃一百道进行介绍。其主要选择原则与呈现风貌有三项:第一,具有比较鲜明的资阳地方特色,包括烹饪技法、风味,菜点创制时间较长;二是采用本地名特食材制作而成,具有从农田到餐桌、一二三产业联动的作用,助推乡村振兴和脱贫攻坚;三是具有较高的文化和旅游吸引价值,传统与创新兼顾,大气与地气兼有,代表性与常见性兼备,能够促进文旅融合,推动资阳经济和社会发展,满足人们美好生活需要。

壹
冷菜

Episode One:
Cold Dishes

01 花椒兔

SICHUAN PEPPER RABBIT

食材配方

兔肉500克　　干青花椒100克
食盐20克　　　味精2克
鸡精2克　　　 辣椒粉30克
姜50克　　　　葱20克
酱油20毫升　　豆豉20克
白糖50克　　　自制香料粉50克
卤水2000毫升

制作工艺

1. 兔肉中加入干青花椒、姜、葱、食盐、酱油、豆豉、白糖、味精、鸡精、辣椒粉、自制香料粉拌匀，腌制2天，捞出晾干。
2. 锅置火上，放入卤水，入兔肉卤熟后捞出、晾凉，斩成条后装盘。

评鉴

兔肉属于高蛋白质、低脂肪、低胆固醇的肉类，兼具美容的作用，深受年轻女性的青睐。此菜主要根据其口味爱好研制而成，色泽红润，肉质细嫩，味道麻而咸鲜。

INGREDIENTS

500g rabbit; 100g dried green Sichuan pepper; 20g salt; 2g MSG; 2g chicken essence; 30g ground chilies; 50g ginger; 20g spring onions; 20ml soy sauce; 20g fermented soya beans; 50g sugar; 50g homemade spice powder; 2,000ml spiced broth

PREPARATION

1. Marinate the rabbit for 2 days with dried green Sichuan pepper, ginger, spring onions, salt, soy sauce, fermented soya beans, sugar, MSG, chicken essence, ground chilies and homemade in spiced broth powder. Remove from the wok and drain.

2. Boil the rabbit with broth in the wok. After the meat is cooked through, remove and drain. Serve at once after cutting into strips.

NOTES

The rabbit is rich in protein and low in fat and cholestenone. It is said that having it makes people more beautiful, so it is favored by young ladies. This dish is reddish, tender, salty and spicy.

02 水晶柠檬 CRYSTAL LEMON

食材配方

鲜柠檬650克　　白糖200克
清水200毫升　　蜂蜜50克
食盐15克

制作工艺

1. 柠檬切片去籽，用清水浸泡4次。
2. 锅置火上，入清水、白糖，用小火熬至浓稠，加入食盐后将锅端离火源，使糖水冷却到40℃左右时加入蜂蜜、柠檬片，再放置火上，用小火加热；如此反复3次即成。

评鉴

安岳柠檬是当地的特产食材，富含柠檬酸和多种维生素、微量元素，具有极强的美容保健作用。此菜采用蜜汁方法去掉了柠檬的苦涩味，色泽淡黄、晶莹透亮，味道香甜可口。

INGREDIENTS

650g fresh lemons; 200g sugar; 200ml water; 50g honey; 15g salt

PREPARATION

1. Cut the lemons into slices and deseed them, soak in water for 4 times.

2. Heat a wok with water and sugar, simmer until thick, add salt and remove the wok from heat. Cool down the sugar-water mixture to 40°C, then add honey and lemon slices, and simmer again. Redo this process for 3 times then done.

NOTES

Anyue Lemon, a local food specialty, has rich citric acid, multi-vitamins and trace elements. It is extremely good for beauty and health maintenance. This dish uses honey to get rid of the bitter taste of lemons. It has slightly yellow color, crystal and lustrous look, sweet and pleasant taste.

食材配方

鸡肉500克　　　姜片20克
葱段50克　　　料酒20毫升
小米辣粒50克　香菜梗粒20克
韭黄碎10克　　姜米20克
蒜米30克　　　白糖10克
复制酱油10毫升　醋30毫升
味精2克　　　　鸡精2克
葱花30克　　　食盐30克
鸡汤100毫升

制作工艺

1 锅置火上，加入鸡肉、姜片、葱段、料酒、食盐，用小火煮至鸡肉八成熟，端离火源，将鸡肉浸泡至熟，捞出晾凉，斩成长约6厘米、粗约1.5厘米的条，装盘。

2 将小米辣粒、香菜梗粒、韭黄碎、姜米、蒜米、白糖、复制酱油、醋、味精、鸡精、鸡汤入碗调成味汁，淋在鸡肉上，撒上葱花即成。

评鉴

资阳酸酸鸡在20年前由乡镇上率先推出，后传入资阳城区，是当地独有菜品，深受民众喜爱，其特点是鸡肉质地细嫩，味道酸辣鲜香、清爽不腻。

ZIYANG SUANSUAN CHICKEN 03 资阳酸酸鸡

INGREDIENTS

500g chicken; 20g ginger, sliced; 50g spring onions, segmented; 20ml Shaoxing cooking wine; 50g bird's eye chilies, finely chopped; 20g coriander, finely chopped; 10g yellow Chinese leeks, finely chopped; 20g ginger, finely chopped; 30g garlic, finely chopped; 10g sugar; 10ml seasoned soy sauce; 30ml vinegar; 2g MSG; 2g chicken essence; 30g spring onions, finely chopped; 30g salt; 100ml chicken broth

PREPARATION

1. Heat the pot, add chicken, ginger slices, spring onions, Shaoxing cooking wine and salt, and simmer until medium well. Remove from heat and wait until the chicken cooked through by the rest heat. Ladle out and cool the chicken. Cut the chicken into about 6cm long and 1.5cm thick strips and transfer to the serving dish.

2. Put the bird's eye chilies, coriander, yellow Chinese leeks, ginger, garlic, sugar, seasoned soy sauce, vinegar, MSG, chicken essence and chicken broth into a bowl and mix them well to make the seasoning sauce. Pour over the sauce on the chicken and sprinkle the spring onions.

NOTES

Ziyang Suansuan Chicken, sour chicken if translated literally, appeared first in rural areas twenty years ago and then had been spread to the downtown areas of Ziyang. It is a popular local specialty which features tender, delicate chicken and sour-spicy flavor.

04 AMBER BEEF SLIVERS

食材配方

牛腿肉500克
干辣椒节15克
花椒5克
八角5克
山柰5克
草果5克
香叶5克
糖色20克
食盐5克
料酒20毫升
芝麻油3毫升
鲜汤250毫升
食用油1 000毫升（约耗100毫升）

制作工艺

1 牛腿肉切成二粗丝，加入食盐、料酒码味10分钟，入180℃的油锅中炸至外酥内熟时捞出。

2 锅置火上，放油烧至120℃，放入干辣椒节、花椒、八角、山柰、草果、香叶炒香，掺入鲜汤，入糖色、食盐、料酒、牛肉丝，用中火收至汁将干时加入芝麻油，出锅晾凉，装盘成菜。

评鉴

此菜色泽似琥珀，酥香化渣，麻辣咸鲜，为佐酒佳肴。

INGREDIENTS

500g silverside; 15g dried chilies, segmented; 5g Sichuan pepper; 5g star anise; 5g sand ginger; 5g caoguo herb; 5g bay leaves; 20g caramel; 5g salt; 20ml Shaoxing cooking wine; 3ml sesame oil; 250ml stock; 1,000ml cooking oil (about 100ml to be consumed)

PREPARATION

1. Cut the silverside into medium slivers and season with salt and Shaoxing cooking wine for 10 minutes. Heat the oil to 180°C and add the slivers, deep fry till the silverside is cooked through inside and crispy outside.

2. Heat the oil to 120°C and add dried chilies segments, Sichuan pepper, star anise, sand ginger, caoguo herb and bay leaves, Stir till aromatic and add stock, caramel, salt, Shaoxing cooking wine and silverside. Simmer over a medium flame till the sauce is reduced and add sesame oil, then transfer to the serving dish.

NOTES

The dish looks like an amber in color and luster and is meltingly crispy and aromatic in taste. It has mala and salty-savory flavors, which is the best choice when drinking wine.

食材配方

鲜藕500克　辣椒油40毫升
红糖25克　　冰糖50克
蚝油20毫升　生抽16毫升
苹果醋90毫升　食用油1000毫升（约耗50毫升）

制作工艺

1. 鲜藕去皮，切成约0.2厘米厚的片，入清水浸泡20分钟，去掉表面淀粉，捞出、沥干水分。
2. 锅置火上，放油烧至150℃，放入藕片炸至表面金黄、酥脆时捞出。
3. 锅置火上，放入辣椒油、红糖、冰糖、蚝油、生抽，用小火熬至冰糖溶化一半时，倒入一半苹果醋，用大火熬至冰糖全部溶化，倒入剩下的苹果醋搅匀，倒出、晾冷，制成糖汁。
4. 将藕片放入糖汁中裹匀，装盘即成。

评鉴

资阳许多地区出产莲藕，尤其以雁江区丹山镇为盛，当地民众善于将莲藕制成多种菜肴，荷塘脆藕就是其中之一，口感酥脆，味道酸甜微辣。

INGREDIENTS

500g fresh lotus roots; 40ml chili oil; 25g brown sugar; 50g rock sugar; 20ml oyster sauce; 16ml light soy sauce; 90ml apple vinegar; 1,000ml cooking oil (about 50ml to be consumed).

PREPARATION

1. Peel the fresh lotus roots and cut into slices 0.2cm thick. Soak in clean water for 20 minutes to remove the starch, and drain.
2. Heat the cooking oil in the wok to 150℃. Deep fry the roots till golden yellow and crispy and remove from the wok.

05 荷塘脆藕
CRISPY POND LOTUS ROOTS

3. Blend in chili oil, brown sugar, rock sugar, oyster sauce and light soy sauce in the wok and simmer till half of the rock sugar melts. Add half of the apple vinegar and boil with high heat till the whole rock candy melts. Add the rest of the apple vinegar and mix well. Remove from the wok and leave to cool till sugar juice is formed.

4. Serve the dish after rolling the lotus root slices in the sugar juice.

NOTES

Lotus roots are growing in many areas in Ziyang City, especially in Danshan Town of Yanjiang District. Locals are good at making lotus roots into different dishes and pond lotus roots are locals' favorite. Crispy Pond Lotus Roots tastes crispy, spicy, sour and sweet.

熊老九卤鹅
06 XIONGLAOJIU SPICED GOOSE

食材配方

鹅1只（约1 500克） 卤鸡蛋2个
卤水2 000毫升 清水500毫升
冰糖500克 白酒250毫升
食用油2 000毫升（约耗30毫升）

制作工艺

1 将鹅用清水浸泡2小时，焯水后捞出，晾干水分，入卤水锅中卤制3小时后捞出，沥干水分；卤鸡蛋去壳。

2 将清水、冰糖、白酒制成混合液，放入卤鹅浸泡2~3小时，捞出、晾干。

3 锅置火上，下油烧至180℃，放入卤鹅炸至酥香后捞出，斩成小块，装盘即成。

评鉴

此菜是用10余种辛香料熬制的卤水卤制而成，色泽棕黄，质地酥软，味道咸鲜香浓。

INGREDIENTS

1 goose（about 1.5kg）; 2 eggs cooked in spiced broth; 2,000ml spiced broth; 500ml water; 500g rock sugar; 250ml liquor; 2,000ml cooking oil (about 30ml to be consumed)

PREPARATION

1. Soak the goose in water for two hours, blanch and drip it to dry. Put the goose in the spiced broth and stew it for 3 hours, then remove and drip it to dry. Put the eggs in the brine and make them well stewed, remove the eggshell.

2. Put water, rock sugar and liquor in a pot and mix well, put the stewed goose in and soak for 2 to 3 hours, then remove and drip to dry.

3. Heat the oil to 180℃. Put the goose in and deep-fry till it is cooked through and crispy. Cut it into small chunks and transfer to the serving dish.

NOTES

This dish is stewed in the broth made of more than ten spices. Its color is brownish yellow and it is tender, crispy, salty and delicious in taste with a strong delicious aroma.

07 山椒泡藕带
PICKLED TENDER LOTUS ROOTS WITH MOUNTAIN PEPPERS

食材配方

藕带300克　　　野山椒50克
野山椒水50毫升　小米辣20克
食盐15克　　　　凉开水400毫升

制作工艺

1. 藕带切成小节，入沸水中焯水后捞出晾凉。
2. 野山椒、野山椒水、小米辣、食盐、凉开水入盆，放入藕带泡制2小时至入味，捞出装盘即成。

评鉴

藕带是莲的幼嫩根状茎，藕带膨大后即成藕。藕带色白、质脆，最适合泡制，成菜酸辣咸鲜，脆嫩爽口，是佐酒佳品。

INGREDIENTS

300g tender lotus roots; 50g mountain peppers; 50ml mountain pepper juice; 20g bird's eye chilies; 15g salt; 400ml cold boiled water

PREPARATION

1. Cut the lotus roots into small segments and blanch them in the boiling water. Then remove and cool.

2. Put mountain peppers, mountain pepper juice, bird's eye chilies, salt and cold boiled water into a pot and then add the lotus roots. Mix and pickle all the ingredients for 2 hours to be tasty and transfer to the serving dish.

NOTES

Tender lotus roots are white in color and crispy in taste, most suitable for pickling. This dish tastes sour, spicy, salty and savory in flavor and crispy, tender and refreshing in taste, which is the best choice when drinking wine.

食材配方

土鸡1只（约1 250克）　食盐50克

花椒20克　味精3克

白糖10克　白酒10毫升

姜片20克　辛香料5克

清水200毫升

制作工艺

1. 用牙签在鸡腿、鸡胸处插一些小孔，放入食盐、花椒、味精、白糖、白酒、姜片、辛香料、清水拌匀，腌制48小时（中途翻面）。

2. 将腌制好的鸡取出，挂通风处晾干，入笼蒸1小时取出，晾干后撕成丝，装盘即成。

评鉴

此菜因制作过程中需将鸡挂于通风处晾干而得名。成菜味道咸鲜香麻，入口干香，佐酒尤佳。

INGREDIENTS

1 free-range chicken (about 1.25kg); 50g salt; 20g Sichuan pepper; 3g MSG; 10g sugar; 10ml liquor; 20g ginger, sliced; 5g spice; 200ml water

PREPARATION

1. Prick the chicken legs and chest, Marinate the chicken for 48 hours in the mixture of salt, Sichuan pepper, MSG, sugar, liquor, ginger, spice and water, and turn occasionlly during the process.

2. Place the pickled chicken in a cool and dry place until dry. Steam for 1 hour, and leave to cool. Shred with hands into slivers and serve.

NOTES

Wind Dried Hard Shredded Chicken is famous for the cooking method. It tastes fresh and salty and more delicious with wine.

08 风吹手撕鸡
WIND DRIED HAND-SHREDDED CHICKEN

09 凉拌仔姜
TENDER GINGER SALAD

食材配方

仔姜400克　食盐5克
白醋5毫升　白糖30克
柠檬汁20毫升　味精2克

制作工艺

1. 将食盐、白醋、白糖、柠檬汁、味精入碗调成味汁。
2. 仔姜切成细丝，加入调味汁拌匀，入冰箱冷藏1小时后装盘即成。

评鉴

此菜选用安岳鼎新乡出产的仔姜。鼎新仔姜，又名紫竹姜，安岳县著名特产，芽长筋少、质地脆嫩、味道辛辣中带有清香。此菜在凉拌时注重突出本味，具有质地脆嫩、味道酸甜、开胃生津的特点。

INGREDIENTS

400g tender ginger; 5g salt; 5ml white vinegar; 30g sugar; 20ml lemon juice; 2g MSG

PREPARATION

1. Make the sauce with salt, white vinegar, sugar, lemon juice and MSG.

2. Cut the tender ginger into slivers, add the sauce and mix well. Place the mix in the refrigerator for 1 hour.

NOTES

The cook uses the tender ginger produced in Dingxin Township, Anyue County. Dingxin tender ginger, which Anyue County is famous for, is also known as Zizhu ginger. It has long buds, few fibres and a pungent and fresh flavors. This dish presents the original taste and crispness of ginger and appetizing sweetness and sourness.

10 凉拌桑叶
MULBERRY LEAVES SALAD

食材配方
食用桑叶200克
食盐5克
小米辣10克
蒜米15克
酱油5毫升
醋8毫升
味精2克
白糖3克
芝麻油5毫升

制作工艺
1. 桑叶入沸水锅中焯水后捞出，晾凉后切碎；小米辣剁碎。
2. 将食盐、小米辣、蒜米、酱油、醋、味精、白糖、芝麻油调匀成酸辣味汁。
3. 将桑叶碎装入圆柱形模具中定型后取出，放入盘中，配上酸辣味汁即成。

评鉴
桑叶具有清肺、明目、降血糖等作用。此菜采用凉拌方式制成，最大限度保持了桑叶的清香，且酸辣可口。

INGREDIENTS
200g edible mulberry leaves; 5g salt; 10g bird's eye chilies; 15g garlic, chopped; 5ml soy sauce; 8ml vinegar; 2g MSG; 3g sugar; 5ml sesame oil

PREPARATION
1. Blanch the mulberry leaves in boiling water, remove and cool down, then chop. Chop the bird's eye chilies.
2. Mix the salt, bird's eye chilies, garlic, soy sauce, vinegar, MSG, sugar and sesame oil and blend well into sour-spicy flavor sauce.
3. Add the chopped mulberry leaves into a cylinder mold to set a shape, then remove and transfer to a serving dish, drizzle with the sour-spicy flavor sauce then serve.

NOTES
Mulberry leaves have benefits for our human bodies, such as eyes, lungs and blood sugar. This dish keeps the utmost natural fragrance of mulberry leaves by using salad cooking style, while it still has sour and pungent taste.

Episode Two: Hot Dishes

食材配方

大鲫鱼1尾（约750克）　茄子100克
碎肉100克　　　　　　面粉300克
糯米粉50克　　　　　　生粉50克
苏打粉2克　　　　　　 纯净水500毫升
干粉丝50克　　　　　　辣椒粉15克
花椒粉5克　　　　　　 食盐2克
料酒12毫升　　　　　　水淀粉20克
食用油2 000毫升（约耗100毫升）

制作工艺

1 将鲫鱼两边的肉片下，保持鱼骨完整；将鱼肉片改刀为约5厘米见方的片，加食盐、料酒码味；碎肉加入食盐、料酒、水淀粉拌匀，制成肉馅；茄子切成连夹片，包入馅肉，制成茄盒初坯。

2 将面粉、糯米粉、生粉、苏打粉、食用油、纯净水调成脆浆糊。

3 锅置火上，放油烧至160℃，放入干粉丝炸至膨松酥脆后捞出，放入盘中垫底；再放入扑有生粉的鱼骨，炸至金黄色时捞出，放在粉丝上。

4 锅置火上，放油烧至150℃，分别将鱼片、茄盒初坯挂上脆浆糊后入锅炸至金黄色后捞出，摆放在盘中鱼骨旁，撒上辣椒粉、花椒粉即成。

评鉴

此菜选用资阳本地产的鲫鱼为主料，结合苌弘故事创新而成。成菜色泽金黄，外酥内嫩，咸鲜香美。

INGREDIENTS

1 carp (about 750g); 100g eggplant; 100g pork mince; 300g wheat flour; 50g glutinous rice flour; 50g starch; 2g baking soda; 500ml pure water; 50g vermicelli; 15g ground chilies; 5g ground Sichuan pepper; 2g salt; 12ml Shaoxing cooking wine; 20g batter; 2,000ml cooking oil (about 100ml to be consumed)

PREPARATION

1. Debone the fish. Cut the fish into about 5cm slices, and marinate with salt and Shaoxing cooking wine. Combine the mince, salt, Shaoxing cooking wine and batter, and blend well to make the stuffing. Cut the eggplant into a pairs of connected slices, and put the stuffing in between.

2. Mix the wheat flour, glutinous rice flour, starch, baking soda, cooking oil and water, whisk well to make the crispiness paste.

3. Heat oil in a wok to 160℃, deep fry the vermicelli till puffy and crispy, and transfer to a serving dish. Coat the fish bones with starch, deep fry till golden brown, and place on the vermicelli.

4. Heat oil in a wok to 150℃, coat the fish slices and stuffed eggplant with crispiness paste, and deep fry till golden brown and cooked through. Place the fried fish slices and eggplant pies on the fish bones. Sprinkle with ground chilies and ground Sichuan pepper.

NOTES

This creative dish based on stories about Chang Hong uses local carp as the main ingredient, featuring golden brown color, crispy ingredients and aromatic, delicate tastes.

02 苌弘鲶鱼
CHANGHONG CATFISH

食材配方

鲶鱼1尾（约1500克）　仔姜100克
临江寺豆瓣50克　　　小米辣50克
姜米6克　　　　　　 蒜米10克
泡辣椒段30克　　　　香葱段20克
芹菜节50克　　　　　啤酒100毫升
味精3克　　　　　　 鲜汤1000毫升
食用油100毫升

制作工艺

1. 鲶鱼改刀成条。

2. 锅置火上，放油烧至120℃，放入临江寺豆瓣、小米辣、姜米、蒜米、泡辣椒段炒香，待油呈红色时，掺入鲜汤，入鲶鱼、啤酒烧至成熟，再入香葱段、芹菜节、味精烧沸，装盘即成。

评鉴

此菜源于苌弘与孔子一起品尝鱼肴的故事，历史悠久，为苌弘文化宴的代表性菜品。成菜色泽红亮，鱼肉细嫩，咸鲜香辣。

INGREDIENTS

1 catfish (about 1,500g); 100g tender ginger; 50g Linjiangsi chili bean paste; 50g bird's eye chilies; 6g ginger, chopped; 10g garlic, chopped; 30g pickled chili peppers, segmented; 20g spring onions, segmented; 50g celery, segmented; 100ml beer; 3g MSG; 1,000ml stock; 100ml cooking oil

PREPARATION

1. Cut the fish into slivers.

2. Heat the oil in a wok to 120°C, blend in Linjiangsi chili bean paste, bird's eye chilies, ginger, garlic and pickled chili peppers, and stir fry till aromatic. Pour in the stock, add the fish and beer, and braise till the fish is cooked through. Add spring onions, celery and MSG, bring to a boil and transfer to a serving dish.

NOTES

This dish has its origin in a story about Chang Hong and Confucius, and is representative of Changhong Cultural Banquet. It features reddish brown color and tender, spicy fish.

食材配方

鲫鱼5尾（约500克）　食盐6克
料酒10毫升　　　　　姜片5克
葱段10克　　　　　　淀粉5克
吉士粉6克　　　　　　米粉6克
泡打粉1克　　　　　　味精2克
鸡精2克　　　　　　　鸡蛋清15克
清水10毫升
食用油1 000毫升（约耗60毫升）

制作工艺

1. 将淀粉、吉士粉、米粉、泡打粉、鸡蛋清、食盐、味精、鸡精、清水调匀成糊。

2. 鲫鱼用食盐、料酒、姜片、葱段码味15分钟，抹上糊，入160℃的油锅中炸至外酥、色呈棕红色，改用小火保持130℃的油温炸至内外酥脆，捞出装盘即成。

评鉴

此菜采用当地出产的鲫鱼制成，色泽自然，内外酥脆，味道咸鲜。

INGREDIENTS

5 crucian carps (about 500g); 6g salt; 10ml Shaoxing cooking wine; 5g ginger, sliced; 10g spring onion, segmented; 5g cornstarch; 6g custard powder; 6g rice powder; 1g baking powder; 2g MSG, 2g chicken essence; 15g egg white; 10ml water; 1,000ml cooking oil (about 60ml to be consumed)

PREPARATION

1. Blend cornstarch, custard powder, rice powder, baking powder, egg white, salt, MSG, chicken essence and water together to make batter.

2. Marinate the carps with salt, Shaoxing cooking wine, ginger and spring onions for 15 minutes. Brush with the batter, put into 160°C oil to deep-fry until reddish brown and crispy, then heat over low flame to keep 130°C to deep-fry until crispy inside, remove and transfer to a serving dish.

NOTES

This dish uses the local crucian carps, and has a fresh, crispy and savory taste.

03 雁江风沙鱼
DEEP-FRIED YANJIANG RIVER CARPS

食材配方

小鲜鲍鱼10个　　鸡腿250克
鸡胗100克　　　杏鲍菇100克
鹌鹑蛋10个　　　食盐6克
桑叶汁15毫升　　金瓜汁50毫升
奶汤200毫升　　 鸡油10克
姜10克　　　　　葱15克
水淀粉30克　　　食用油30毫升

制作工艺

1 将鲍鱼改成"十"字花刀，入沸水锅中焯水后备用；鸡腿宰成长约6厘米、宽约2厘米的条，入180℃的油中炸至色金黄色时捞出；杏鲍菇切成约1厘米厚的片，入160℃的油中炸至色金黄色时捞出；鸡胗切成约1厘米厚的片，焯水后备用；鹌鹑蛋煮熟后去壳，抹上酱油后入油锅炸为金黄色。

2 锅置火上，放油烧至150℃，放入姜、葱炒香，掺入奶汤烧出香味，捞出姜、葱不用，放入鸡腿、鸡胗、杏鲍菇、鹌鹑蛋烧至软糯，放入鲍鱼、食盐烧入味，捞出后装盘。

3 锅中加入原汤、桑叶汁、金瓜汁、鸡油烧沸，加入水淀粉收汁至浓稠，浇淋在菜上即成。

评鉴

此菜荤素搭配，色彩丰富，质地软糯，味道咸鲜，香味浓郁。

INGREDIENTS

10 small fresh abalones; 250g chicken legs; 100g chicken gizzards; 100g king brown mushrooms; 10 quail eggs; 6g salt; 15ml mulberry juice; 50ml pumpkin sauce; 200ml milky stock; 10g chicken fat; 10g ginger; 15g spring onion; 30g batter; 30ml cooking oil

PREPARATION

1. Cut crosses on the abalones, blanch in boiling water. Cut the chicken legs into about 6cm long, 2cm wide strips, then deep-fry in 180℃ oil till golden brown and remove. Cut the king brown mushrooms into about 1cm thick slices, then deep-fry in 160℃ oil till golden brown and remove. Cut the chicken gizzards into 1cm thick slices, blanch. Boil the quail eggs until cooked through, remove the shell, brush with soy sauce, deep-fry in oil till golden brown.

2. Heat oil in a wok to 150℃, add ginger and spring onions to stir-fry till aromatic, pour milky stock to braise till it brings out fragrance, remove the ginger and spring onions. Add the chicken legs, chicken gizzards, king brown mushrooms and quail eggs and braise until the contents are soft and glutinous, add abalones and salt to braise until fully absorb the flavor then remove and transfer to a serving dish.

3. Add the braised stock, mulberry leaves sauce, pumpkin sauce and chicken fat in the wok to heat until boiling, add batter and wait till the sauce thickens and pour over the dish.

NOTES

With rich color of meat and vegetables, this dish is aromatic and has soft and glutinous, salty and savory taste.

04 桑汁烩鲍鱼
BRAISED ABALONES IN MULBERRY JUICE

食材配方

鲜鲍10个	桑芽10克
土鸡250克	玉米棒200克
虫草花20克	干桂圆50克
食盐7克	味精2克
料酒10毫升	矿泉水1000毫升

制作工艺

1. 鲜鲍用食盐码味后用温水浸泡，去壳及内脏。

2. 土鸡斩成约3厘米见方的块，焯水；桑芽用清水浸泡；干桂圆去壳；玉米棒斩成段；虫草花用清水泡软。

3. 将矿泉水倒入紫砂壶中，再放入桑芽、桂圆、鲜鲍、鸡肉块、虫草花、食盐、味精，加盖上笼蒸3.5小时，取出即成。

评鉴

桑芽具有清肺润燥、清肝明目的作用。此菜将桑芽与鲜鲍、土鸡、桂圆、虫草花等食材搭配制成，入口清香，回味甘甜，更具营养和食疗价值。

INGREDIENTS

10 fresh abalones; 10g mulberry sprouts; 250g free-range chicken; 200g corns on the crop; 20g cordyceps flowers; 50g longan; 7g salt; 2g MSG; 10ml Shaoxing cooking wine; 1,000ml mineral water

PREPARATION

1. Smear the fresh abalones with salt and soak in warm water. Remove their shell and viscera.

2. Cut the chicken into about 3cm diamonds and boil to be medium well. Soak the mulberry sprouts in water. Remove the shells of longan. Cut the corns into segments. Soak the cordyceps flowers in water till soft.

3. Pour water into an enameled pottery, and place mulberry sprouts, longan, fresh abalones, chicken, cordyceps flowers, salt and MSG in the pottery and cover it with a cap. Steam for 3.5 hours and transfer to a dish.

NOTES

Mulberry sprouts can help clear the lung heat, moisten dryness, clear liver and improve vision. This dish is made out of mulberry sprouts, chicken, longan, and cordyceps flowers. It tastes fragrant, fresh and sweet. It is nutrient and good for people's health.

06 柠香脆皮鱼
LEMON FLAVOR CRISPY FISH

食材配方

鲤鱼1尾（约800克）	姜片10克
葱段20克	大葱30克
泡辣椒30克	水淀粉20克
姜米10克	蒜米20克
葱花25克	白糖100克
食盐8克	酱油5毫升
醋10毫升	味精2克
柠檬汁25毫升	料酒20毫升
鲜汤100毫升	食用油2 000毫升（约耗150毫升）

制作工艺

1. 鲤鱼宰杀后洗净，在鱼身两侧各剞牡丹花刀；鱼头用刀斩破，加入姜片、葱段、食盐、料酒码味；大葱、泡辣椒分别切成细丝，加清水浸泡。

2. 将食盐、白糖、酱油、醋、味精、柠檬汁、料酒、水淀粉、鲜汤入碗调成芡汁。

3. 锅置火上，放油烧至180℃，将鱼身挂上一层水淀粉糊后入锅炸至定型，捞出备用；待油温回升至200℃时，将鱼再次放入，炸至外酥内熟时捞出，装入盘中。

4. 锅置火上，放油烧至120℃，放入姜米、蒜米、葱花炒香，倒入芡汁，待收汁亮油后浇淋在鱼上，撒上葱丝和泡辣椒丝即成。

评鉴

此菜在糖醋脆皮鱼的基础上加入柠檬汁，酸味更为丰富，成菜具有形态美观、外酥内嫩、味道酸甜香醇的特点。

INGREDIENTS

1 Carp(about 800g); 10g ginger slices; 20g spring onions, segmented; 30g spring onions; 30g pickled chilies; 20g batter; 10g ginger, finely chopped; 20g garlic, finely chopped; 25g spring onions, finely chopped; 100g sugar; 8g salt; 5ml soy sauce; 10ml vinegar; 2g MSG; 25ml lemon juice; 20ml Shaoxing cooking wine; 100ml stock; 2,000ml cooking oil (about 150ml to be consumed)

PREPARATION

1. Kill and rinse the carp. Cut 2 inches long and 0.5 inches deep, and then cut backward slightly to fish head side to have connected slices, and cut 1 or 2 times on inside flesh. Redo this cutting every 2 inches wide on both fish sides. Lift the fish by the fish tail, and it looks like a flower. Cut the fish head into half, marinate with ginger, spring onions, salt and Shaoxing cooking wine. Cut spring onions and pickled chilies into slivers separately, and soak in water.

2. Mix salt, sugar, soy sauce, vinegar, MSG, lemon juice, Shaoxing cooking wine, batter and stock to make the thickening sauce.

3. Heat oil in a wok to 180°C, coat the fish with batter, then put into the wok to fry for one or two minutes and remove. Heat the oil to 200°C, deep-fry the fish for a second time till the skin becomes brown, crispy and the fish is cooked through, then transfer to a serving dish.

4. Heat some oil in a wok to 120°C, add chopped ginger, garlic, spring onions to fry until aromatic. Add the thickening sauce until it becomes thicken, then pour it over the fish. Sprinkle with pickled chili slivers and spring onion slivers.

NOTES

This dish adopts from the dish Sweet-and-sour Crispy Fish. It has stronger sour taste after adding some lemon juice. This dish has nice look, tender meat but crispy skin, rich sour and savory taste.

07 大蒜豆瓣鲜鱼
FISH IN GARLIC AND CHILI BEAN SAUCE

INGREDIENTS

750g fish; 20 one-clove garlic; 5g ginger, sliced; 10g spring onions, segmented; 50g Linjiangsi chili bean paste; 10g ginger, finely chopped; 15g garlic, finely chopped; 30g spring onions, finely chopped; 10ml soybean sauce; 20g sugar; 15ml vinegar; 25ml Shaoxing cooking wine; 2g MSG; 3g salt; 1g pepper; 30g batter; 750ml stock; 150ml cooking oil

PREPARATION

1. Score five slashes on both sides of the fish, season with salt, Shaoxing cooking wine, ginger slices, spring onions and pepper.

2. Heat the oil to 200℃ and deep fry the fish till both sides are golden; Remove the fish and add the one-clove garlic in the 150℃ oil, deep fry till the peels are wrinkled.

3. Heat the oil to 120℃ and add Linjiangsi chili bean paste, ginger, garlic and spring onions, stir till aromatic and the oil turns to glistening red color. Add stock, fish, one-clove garlic, soybean sauce, sugar and Shaoxing cooking wine, boil till the fish is cooked through. Transfer the fish to an oval plate and put one-clove garlic around; Add vinegar, MSG, batter and spring

食材配方

鲜鱼750克	独蒜20个
姜片5克	葱段10克
临江寺豆瓣50克	姜米10克
蒜米15克	葱花30克
酱油10毫升	白糖20克
醋15毫升	料酒25毫升
味精2克	食盐3克
胡椒粉1克	水淀粉30克
鲜汤750毫升	食用油150毫升

制作工艺

1. 在鱼身两面各剞5刀，加食盐、料酒、姜片、葱段、胡椒粉码味。

2. 锅置火上，放油烧至200℃，放入鱼炸至两面金黄时捞出；独蒜入150℃油中炸至皱皮后捞出。

3. 锅置火上，放油烧至120℃，放入临江寺豆瓣、姜米、蒜米、葱花炒香，待油呈红色时，掺入鲜汤，入鱼、独蒜、酱油、白糖、料酒，烧至鱼熟入味后捞出装入条盘，独蒜围边；锅内放入醋、味精、水淀粉、葱花，待收汁亮油后浇淋在鱼上即成。

评鉴

此菜在川菜经典菜肴豆瓣鲜鱼的基础上重用大蒜，色泽红亮，鱼肉细嫩，味道咸鲜酸甜，蒜香浓郁。

onions to the wok, fry till the sauce is reduced and glistening, then pour over the fish.

NOTES

The dish adds large amounts of garlic on the basis of Sichuan classic cuisine Fish in Chili Bean Sauce. It is glistening red in color, tender and soft in mouth feel. It tastes salty, delicate, sour, sweet and has a strong aroma of garlic.

食材配方

河鲤1 000克	自制水豆豉250克
五花肉粒100克	小米辣20克
杏鲍菇粒100克	碎米芽菜10克
临江寺豆瓣10克	姜片5克
姜米6克	蒜米10克
葱段10克	味精2克
鸡精2克	酱油10毫升
花椒2克	白糖1克
食盐5克	料酒20毫升
鲜汤750毫升	
食用油1 000毫升（约耗100毫升）	

制作工艺

1. 将河鲤剞"一"字花刀，加食盐、料酒、姜片、葱段码味后，入180℃的油中炸至色棕红时捞出备用。
2. 锅置火上，放油烧至150℃，放入五花肉粒炒香，入临江寺豆瓣、自制水豆豉、花椒、姜米、蒜米、小米辣、杏鲍菇粒、碎米芽菜炒香，掺入鲜汤，入河鲤、酱油、白糖、料酒，烧至鱼熟入味后捞出装入盘中；锅中加入味精、鸡精收至浓稠，浇淋在鱼上成菜。

评鉴

此菜创意源于怀念老家的美味，用当地的特色调味料临江寺豆瓣、自家制作的水豆豉等调出儿时记忆中的味道，地方特色突出，咸鲜微辣，香味浓郁。

INGREDIENTS

1,000g river carps; 250g homemade fermented soybeans; 100g pork belly cubes; 20g bird's eye chilies; 100g king brown mushroom cubes; 10g chopped Yacai; 10g Linjiangsi chili bean paste; 5g ginger, sliced; 6g ginger, chopped; 10g garlic, chopped; 10g spring onion, segmented; 2g MSG; 2g chicken essence; 10ml soy sauce; 2g Sichuan pepper; 1g sugar; 5g salt; 20ml Shaoxing cooking wine; 750ml stock; 1,000ml cooking oil (about 100ml to be consumed)

PREPARATION

1. Cut lines on the carp, add salt, Shaoxing cooking wine, gingers, spring onions to marinate. Heat oil in a wok to 180°C, add the fish to deep-fry until reddish brown then remove.

2. Heat oil in a wok to 150°C, add pork cubes to stir-fry until aromatic, add Linjiangsi chili bean paste, homemade fermented soybeans, Sichuan pepper, ginger, garlic, bird's eye chilies, king brown mushrooms, Chopped Yacai to stir-fry to bring out the aroma. Pour in the stock, add the fish, soy sauce, sugar, Shaoxing cooking wine to braise until cooked through, remove and transfer to a serving dish. Add MSG and chicken essence in the stock until it becomes thick, then pour over the fish and serve.

NOTES

Some ingredients like Linjiangsi chili bean paste and homemade fermented soybeans can remind us about our childhood memory of hometown food. With savoury, aromatic and slightly spicy taste, this dish has unique regional features.

BRAISED TUO RIVER CARPS 08 沱江河鲤

09 船老板鱼
FISHERMAN'S FISH

食材配方

鲜鱼800克　　　猪五花肉100克
泡菜100克　　　泡姜50克
泡辣椒段50克　　葱段50克
蒜50克　　　　　芹菜节40克
食盐5克　　　　　味精3克
白糖3克　　　　　料酒15毫升
食用油150毫升　　鲜汤1000毫升

制作工艺

1. 鲜鱼切成块；猪五花肉切成小片；泡菜、泡姜分别切成片。
2. 锅置火上，放油烧至180℃，放入猪五花肉炒香，入泡菜、泡姜、泡辣椒段、葱段、蒜炒香，掺入鲜汤，放入鱼块、食盐、味精、白糖、料酒、芹菜节，用中火烧至鱼熟后装盘即成。

评鉴

此菜源于沱江打鱼人家烧鱼的烹调方法，后经改进而成，鱼肉细嫩鲜美、味道咸鲜略带酸辣。

INGREDIENTS

800g fish; 100g pork belly; 100g pickles; 50g pickled ginger; 50g pickled chilies; 50g spring onions, segmented; 50g garlic; 40g celery, segmented; 5g salt; 3g MSG; 3g sugar; 15ml Shaoxing cooking wine; 150ml cooking oil; 1,000ml stock

PREPARATION

1. Chunk the fish, and cut the pork belly into small slices. Slice the pickles and pickled ginger.

2. Heat the oil in a wok to 180°C, add the pork belly, and stir fry till aromatic. Blend in the pickles, pickled ginger, pickled chilies, spring onions and garlic, and continue to stir to bring out the aromas. Add the stock, fish, salt, MSG, sugar, Shaoxing cooking wine, celery, and braise over medium heat till the fish is cooked through. Transfer to a serving dish.

NOTES

The dish is a modern twist of a traditional fisherman dish on the Tuojiang River, featuring tender, salty and slightly sour and spicy fish.

10 剁椒沱江鱼头煲

BRAISED TUOJIANG FISH HEAD WITH CHOPPED CHILIES

食材配方

鳙鱼头1 000克　自制泡菜酱150克
红剁椒酱50克　　黄剁椒酱50克
芋头350克　　　食盐10克
姜片15克　　　　葱段20克
料酒15毫升　　　鸡油20克
鲜汤200毫升

制作工艺

1 鱼头剖开成两半，加入食盐、姜片、葱段、料酒腌制20分钟；芋头切块后煮熟备用。

2 芋头块入煲仔垫底，放上鱼头，加入鲜汤、鸡油，在鱼头上放入自制泡菜酱，再加入红、黄两种剁椒酱，加盖，用旺火煲5分钟后改用小火煲4分钟至熟即成。

评鉴

沱江河里的鳙鱼肉质鲜美，大鳙鱼头最适合用剁椒来蒸制。此菜汤汁浓稠，鱼头味道辣酸鲜香。

INGREDIENTS

1,000g bighead carp head; 150g home-made pickled vegetable sauce; 50g chopped red chili sauce; 50g chopped yellow chili sauce; 350g taros; 10g salt; 15g ginger, sliced; 20g spring onions, segmented; 15ml Shaoxing cooking wine; 20g chicken fat; 200ml stock

PREPARATION

1. Cut the fish head into halves and season with salt, ginger slices, spring onions segments and Shaoxing cooking wine for 20 minutes; cut the taros into chunks and boil till they are cooked through.

2. Put the taros at the bottom of the pot and then put the fish head in the pot, add stock, chicken fat, home-made pickled vegetable sauce, red chili sauce and yellow chili sauce, cover the pot with its lid and braise over a high flame for 5 minutes and then turn to a low flame for another 4 minutes until the fish head is cooked through.

NOTES

The bighead carp in Tuojiang are noted for their delicious and nutritious meat. The carp head is best for braising with chopped chilies. This dish has thick sauce with sour, spicy, savory and dainty tastes.

食材配方

草鱼1尾（约1 500克）　桑葚汁100克
番茄酱50克　　　　　　姜片5克
葱段15克　　　　　　　食盐6克
白糖100克　　　　　　 白醋50毫升
料酒5毫升　　　　　　 淀粉150克
水淀粉30克　　　　　　清水100毫升
食用油1 000毫升（约耗60毫升）

INGREDIENTS

1 grass carp (about 1,500g); 100g mulberry juice; 50g ketchup; 5g ginger, sliced; 15g spring onions, segmented; 6g salt; 100g sugar; 50ml white vinegar; 5ml Shaoxing cooking wine; 150g cornstarch; 30g batter; 100ml water; 1,000ml cooking oil (about 60ml to be consumed)

制作工艺

1. 草鱼去头、尾和骨，取净鱼肉先剞成"十"字花刀，再切成三角块，加入姜片、葱段、食盐、料酒码味10分钟后均匀地粘上淀粉。

2. 锅置火上，放油烧至160℃，入鱼块炸至成熟、呈菊花形时捞出；待油温回升到180℃，放入鱼块炸至表面呈金黄色后捞出装盘。

3. 锅置火上，放油烧至120℃，放入番茄酱炒香，掺入清水，入桑葚汁、白糖、白醋、水淀粉勾汁成二流芡，浇淋在鱼上即成。

评鉴

桑葚是一种药食同源的食品，有提高免疫力、修复肝肾功能的作用，号称"赛人参"。此菜在菊花鱼的基础上加入桑葚，不但形态美观，还兼具一定的食疗保健作用。

11 桑葚菊花鱼
CHRYSANTHEMUM FISH IN MULBERRY SAUCE

PREPARATION

1. Remove the head and tail of the grass carp, and debone to get fish fillets. Cut crosses on the fillets, then slice into triangle pieces, add ginger, spring onions, salt, Shaoxing cooking wine to marinate for 10 minutes, then coat with cornstarch evenly.

2. Heat oil in a wok to 160°C, add the fish fillets to deep-fry until well-cooked and shape like a chrysanthemum then ladle out.

3. Heat oil in a wok to 120°C, add ketchup to stir-fry until aromatic, pour in water. Add mulberry juice, sugar, white vinegar, batter into a seasoning sauce, drizzle over the fish then done.

NOTES

Mulberry is not only a fruit but also has benefits to our health. It has a nickname "fruit ginseng", which benefits our immune system and body organs like liver and kidney. This dish add mulberry on the basis of Deep-fried Chrysanthemum Fish. It is a successfully combination of a pleasant look and health benefits in one dish.

12 家常沱江翘壳
HOMEMADE TUOJIANG QIAOKE CARP

食材配方

翘壳鱼1尾（约1000克）　仔姜丝100克
小米辣丝75克　葱段50克
临江寺豆瓣25克　泡辣椒末100克
姜米25克　蒜米50克
干青花椒25克　干红花椒10克
食盐10克　啤酒100毫升
胡椒粉2克　白糖10克
醋20毫升　醪糟汁15毫升
鸡精2克　味精2克
葱花50克　红薯粉50克
清水1000毫升　水淀粉50克
食用油200毫升

制作工艺

1 翘壳鱼斩成条，加入食盐、啤酒浸泡10分钟后用清水洗净，加入红薯粉拌匀，入油锅定型后捞出。

2 锅置火上，放油烧至120℃，放入临江寺豆瓣、泡辣椒末、姜米、蒜米、葱段、干青花椒、干红花椒炒香，先入醪糟汁、小米椒丝、啤酒、清水烧沸，再入鱼条、胡椒粉、白糖、醋、鸡精、味精烧至成熟，入水淀粉勾成薄芡，装盘后撒上葱花即成。

评鉴

翘壳鱼，学名白鱼，因嘴部上翘而得名，肉白而细嫩，味道鲜美，是一种食用价值非常高的鱼类，多用于家常烧制。此菜色泽红亮，鱼肉细嫩，咸鲜香辣。

INGREDIENTS

1 qiaoke carp (about 1,000g); 100g tender ginger slivers; 75g bird's eye chilies slivers; 50g spring onions, segmented; 25g Linjiangsi chili bean paste; 100g pickled pepper, finely chopped; 25g ginger, finely chopped; 50g garlic, finely chopped; 25g dried green Sichuan peppercorns; 10g dried red Sichuan peppercorns; 10g salt; 100ml beer; 2g pepper; 10g sugar; 20ml vinegar; 15ml fermented rice juice; 2g chicken essence; 2g MSG; 50g spring onions, finely chopped; 50g sweet potato starch; 1,000ml water; 50g batter; 200ml cooking oil

PREPARATION

1. Cut the fish into strips. Soak in salt and beer for 10 minutes. Wash the strips clean and mix with sweet potato starch. Fry in the wok till the shape is set and remove from the work.

2. Heat the oil in the wok till 120°C. Stir fry Linjiangsi chili bean paste, pickled pepper, ginger, garlic spring onions, dried green Sichuan peppercorns, and dried red Sichuan peppercorns. Blend in the fermented rice juice, bird's eye chilies, beer and water and boil. Add the fish, pepper, sugar, vinegar, chicken essence and MSG and boil till cooked through. Add batter to thicken the soup and serve at once with spring onions.

NOTES

A qiaoke carp has an up-pointing mouth, hence its name. Its meat is tender and fresh, rich in nutrition, and is enjoyed by households. This dish is brown reddish, salty and spicy.

13 炝锅小麻鱼

SCORCHING STIR-FRIED STONE MOROKO

食材配方

小麻鱼100克	干辣椒碎200克	香菜100克
小米椒100克	葱花50克	姜片10克
姜米40克	蒜米150克	豆豉末50克
花生碎150克	芝麻30克	花椒30克
花椒粉20克	食盐15克	味精2克
鸡精2克	芝麻油10毫升	花椒油10毫升
料酒250毫升	食用油1 000毫升（约耗60毫升）	

制作工艺

1. 将小麻鱼加入食盐、姜片、花椒、料酒、香菜、小米辣拌匀腌渍2小时。
2. 干辣椒入锅，用小火慢慢烤香，取出晾冷后压碎。
3. 锅置火上，放油烧至180℃，放入小麻鱼炸至酥脆、表面呈金黄色时捞出装盘。
4. 锅置火上，放油烧至150℃，放入姜米、蒜米、豆豉末、干辣椒碎炒香，再入味精、鸡精、花椒粉、花生碎、芝麻、花椒油、芝麻油、葱花炒匀，出锅后淋在鱼上即成。

评鉴

资阳境内的沱江鱼类资源丰富，小麻鱼容易捕捞，但因肉少刺多，大多用油炸成菜，口感香辣酥脆，尤其适合佐酒。

INGREDIENTS

100g stone moroko; 200g dried chilies, finely chopped; 100g coriander; 100g bird's eye chilies; 50g spring onions, finely chopped; 10g ginger, sliced; 40g ginger, finely chopped; 150g garlic, finely chopped; 50g fermented soy beans, finely chopped; 150g peanuts, finely chopped; 30g sesames; 30g Sichuan pepper; 20g ground Sichuan pepper; 15g salt; 2g MSG; 2g chicken essence; 10ml sesame oil; 10ml Sichuan pepper oil; 250ml Shaoxing cooking wine; 1,000ml cooking oil (about 60ml to be consumed)

PREPARATION

1. Season the stone moroko with salt, ginger slices, Sichuan pepper, Shaoxing cooking wine, coriander and bird's eye chilies for 2 hours.

2. Put the dried chilies in the wok and roast over a low flame till aromatic, then remove the chilies and finely chop them.

3. Heat the oil to 180°C and deep fry the stone moroko till crispy and golden.

4. Heat the oil to 150°C and add ginger, garlic, fermented soy beans and dried chilies, stir till aromatic. Add MSG, chicken essence, ground Sichuan pepper, peanuts, sesames, Sichuan pepper oil, sesame oil and spring onions, stir and mix them well. Then pour the seasoning over the stone moroko.

NOTES

The Tuojiang River within the Ziyang city of Sichuan boasts abundant fish species, among which stone moroko is easy to be caught. However, stone moroko has a large number of bones and little meat. Therefore, it is often deep fried and tastes spicy, savory and crispy, which is the best choice when drinking wine.

14 白乌鱼炖土鸡
WHITE SNAKEHEAD AND CHICKEN STEW

食材配方

白乌鱼1尾（约600克） 土鸡500克
姜片10克 葱段20克
料酒20毫升 胡椒粉1克
食盐10克 味精2克

制作工艺

1 白乌鱼和土鸡均斩成约4厘米见方的块，入沸水锅中焯水后备用。

2 锅置火上，加入清水，下土鸡、姜片、葱段、料酒、胡椒粉，用大火烧沸后，改用小火炖2小时至鸡肉熟软，然后放入白乌鱼块炖10分钟，入食盐、味精调味，起锅即成。

评鉴

白乌鱼是资阳市乐至县的特产，国家地理标志产品，有鱼中珍品之称，色白、无刺、肉质细嫩，味道鲜美，营养价值极高，具有一定的滋补作用。此菜将白乌鱼与同样具有滋补作用的土鸡同炖，滋补效果更佳。

INGREDIENTS

1 white snakehead (about 600g); 500g free-range chicken; 10g ginger, sliced; 20g spring onions, segmented; 20ml Shaoxing cooking wine; 1g pepper; 10g salt; 2g MSG

PREPARATION

1. Cut the white snakehead and chicken into about 4cm chunks. Blanch the chicken.

2. Heat a wok, add water, chicken, ginger, spring onions, Shaoxing cooking wine and pepper, and bring to a boil. Reduce to low heat and simmer for 2 hours till the chicken is tender and cooked through. Add the fish, and continue to simmer for 10 minutes. Blend in the salt and MSG to season, and remove from the heat.

NOTES

White snakehead, a local specialty and a national geographical indication produce, is a rare species of fish that features a white body, few bones, tender meat, rich nutrition and a delicate taste. This dish, combining the nourishing functions of both snakehead and chicken, is a health benefiting stew.

15 倒罐蒸桑叶鸡
CHICKEN STEAMED WITH MULBERRY LEAVES

食材配方

桑叶土鸡1只（1500克）　食用桑叶100克
白酒6毫升　　姜片15克
葱段30克　　干辣椒节3克
花椒1.5克　　胡椒粉1克
食盐10克　　大枣3个
枸杞10克

制作工艺

1. 将整鸡抹上白酒后搓揉10分钟，再抹上食盐、胡椒粉，腌制40分钟。
2. 将整鸡放入倒罐中，依次加入食用桑叶、姜片、葱段、干辣椒节、花椒、大枣、枸杞，加盖，盖子内装满冷水。
3. 将倒罐置于铁锅上，隔水蒸制6小时（期间需不断更换盖子内的水，水温保持在20℃以内）即成。

评鉴

倒罐蒸土鸡是乐至县回澜镇民间传承两百余年的传统名菜，蒸制时不加一滴水，全靠蒸汽倒流形成汤汁而得名。此菜用桑叶喂养的土鸡加上桑叶等食材蒸制而成，汤汁清澈见底，鸡肉酥软嫩滑，味道咸鲜，香味浓郁。

2. Place the chicken in the crock whose cap is concave and add gouqi berry leaves, ginger, spring onions, dried chilies, Sichuan pepper, dates, and wolfberries in the crock. Cover the crock with a cap and fill the cap with cold water.

3. Place the crock on the wok and steam for 6 hours (keep changing the water in the cap during the period and keep the temperature of the water below 20°C). Serve the dish.

NOTES

Chicken Steamed with Mulberry Leaves enjoys a history of over 200 years and is a traditional dish in Huilan Town of Lezhi County. During the steaming, no water is added, and the chicken is steamed only by the vapor. With wolfberries and local chickens fed with mulberry leaves, the soup is clear and this chicken tastes tender, soft, and delicate.

INGREDIENTS

1,500g chicken fed with mulberry leaves; 100g edible mulberry leaves; 6ml liquor; 15g ginger, sliced; 30g spring onions, segmented; 3g dried chilies, segmented; 1.5g Sichuan pepper; 1g pepper; 10g salt; 3 dates; 10g wolfberries

PREPARATION

1. Smear the whole chicken skin with liquor and rub for 10 minutes. Smear salt and pepper on the skin and marinate for 40 minutes.

16 藿香乌鱼花
DEEP FRIED SNAKEHEAD WITH HUOXIANG HERB

食材配方

乌鱼750克	鸡蛋清50克
淀粉50克	藿香碎100克
姜片5克	葱段10克
葱花50克	小米辣50克
仔姜50克	海鲜酱油20毫升
黄豆酱油10毫升	料酒10毫升
食盐4克	鸡精2克
味精2克	藤椒油10毫升
清汤750毫升	
食用油1 000毫升（约耗50毫升）	

制作工艺

1. 乌鱼取净肉，先刨成"十"字花刀，再切成约5厘米见方的块，加入食盐、姜片、葱段、料酒码味10分钟。

2. 鸡蛋清与淀粉拌匀成蛋清淀粉；小米辣、仔姜切成颗粒。

3. 锅置火上，放油烧至150℃，将乌鱼花与蛋清淀粉拌匀后入锅炸熟后捞出，摆放在盘中。

4. 锅置火上，掺入清汤，放入小米辣粒、仔姜粒、海鲜酱油、黄豆酱油、鸡精、味精、藤椒油烧沸，制成味汁，起锅后浇淋在乌鱼花上，撒上葱花、藿香碎即成。

评鉴

此菜因调味重用藿香而得名，鱼肉质地Q弹细滑，咸鲜麻辣，藿香味突出。

INGREDIENTS

750g snakehead; 50g egg white; 50g cornstarch; 100g huoxiang herb, finely chopped; 5g ginger, sliced; 10g spring onions, segmented; 50g spring onions, chopped; 50g bird's eye chilies; 50g tender ginger; 20ml seafood sauce; 10ml soy sauce; 10ml Shaoxing cooking wine; 4g salt; 2g chicken essence; 2g MSG; 10ml pepper oil; 750ml consommé; 1,000ml cooking oil (about 50ml to be consumed)

PREPARATION

1. Debone the fish, excert cross cuts into the flesh, and cut into 5cm sections. Marinate the fish with salt, ginger slices, garlic segments and Shaoxing cooking wine for 10 minutes.

2. Mix the egg white with cornstarch. Cut the bird's eye chilies and tender gingers into grains.

3. Heat the cooking oil in the wok to 150°C. Mix the fish with egg white and fry till fried through. Remove from the wok and place in the plate.

4. Add consommé, bird's eye chilies, ginger, seafood sauce, soy sauce, chicken essence, MSG, pepper oil and heat till they are boiling to form sauce. Sprinkle the sauce on the fish. Serve at once with spring onions and huoxiang herb.

NOTES

Deep Fried Snakehead with Huoxiang Herb is famous for the flavor of huoxiang. The fish is tender, salty and spicy with a strong flavor of huoxiang.

17 干烧乌鱼片
DRY-BRAISED SNAKEHEAD SLICES

食材配方

乌鱼600克　　青尖椒30克
泡辣椒20克　　姜米5克
蒜片8克　　　葱段20克
食盐5克　　　料酒10毫升
鸡蛋清30克　　淀粉30克
味精2克　　　芝麻油3毫升
鲜汤100毫升　食用油1 000毫升（约耗70毫升）

制作工艺

1. 乌鱼去头、去骨，片成约0.3厘米厚的片，加入食盐、料酒、鸡蛋清、淀粉拌匀；泡辣椒、青尖椒分别切成菱形块。
2. 锅置火上，放油烧至160℃，放入乌鱼片炸至定型后捞出。
3. 锅置火上，放油烧至120℃，放入泡辣椒块、青尖椒块、姜米、蒜片、葱段炒香，掺入鲜汤烧沸，入食盐、料酒、乌鱼片，烧至水分将干时，入味精、芝麻油搅匀，起锅装盘即成。

评鉴

资阳境内江河纵横，盛产各种鱼类，乌鱼是其中代表性特色产品。此菜用干烧的方法烹制而成，色泽红亮、鱼肉细嫩、味道咸鲜微辣。

INGREDIENTS

600g snakehead; 30g green chilies; 20g pickled chilies; 5g ginger, finely chopped; 8g garlic sliced; 20g spring onions, segmented; 5g salt; 10ml Shaoxing cooking wine; 30g egg white; 30g cornstarch; 2g MSG; 3ml sesame oil; 100ml stock; 1,000ml cooking oil (about 70ml to be consumed)

PREPARATION

1. Remove the head and bones of the snakehead, and cut into about 0.3cm slices. Add salt, Shaoxing cooking wine, egg white and cornstarch and then mix them well. Cut pickled chilies and green chilies into rhombus chunks.

2. Heat the oil to 160℃ in a wok and deep fry the fish slices until they are shaped, and then remove from the oil.

3. Heat the oil to 120℃ in a wok and add pickled chilies, green chilies, ginger and garlic to fry till aromatic. Then add the stock, bring to a boil and add salt, Shaoxing cooking wine and fish slices to braise until the water is nearly dried. Blend in MSG and sesames and transfer to the serving dish.

NOTES

Ziyang city boasts many rivers and lakes, producing many kinds of fish, of which the most famous is snakehead. This dish is cooked by using dry braising. Its color is bright red and it tastes tender, soft, fresh and a little spicy.

FISH WITH PORK INTESTINES
18 肥肠鱼

食材配方

花鲢鱼2 500克	熟肥肠500克
干辣椒15克	花椒5克
姜片10克	大蒜20克
蒜米20克	葱花40克
芹菜段30克	泡菜50克
芝麻3克	临江寺豆瓣40克
秘制香料30克	食盐8克
味精2克	料酒15毫升
鲜汤3 000毫升	食用油80毫升

制作工艺

1 花鲢鱼取净肉片成片，鱼骨斩成段，分别用食盐、料酒、味精拌匀腌制。

2 锅置火上，放油烧至100℃，放入临江寺豆瓣、花椒、熟肥肠、泡菜、葱段、姜片、大蒜炒出香味，再入秘制香料、鲜汤烧沸，入鱼头、鱼骨煮熟后捞出，装入盆中；锅中再放入鱼片，用小火煮熟，倒入盆中，放上芹菜段、葱花、蒜米。

3 另锅置火上，放油烧至150℃，放入干辣椒、花椒、芝麻炒香，浇淋在盆中即成。

评鉴

资阳河流纵横，鱼鲜丰富。肥肠鱼在资阳市许多地方都有制作，色泽红亮，味道麻辣浓香，鱼片鲜嫩，肥肠弹性足，价廉物美，乡土气息浓郁。

INGREDIENTS

2,500g spotted silver carp; 500g cooked intestine; 15g dried chili peppers; 5g Sichuan pepper; 10g ginger, sliced; 20g garlic; 20g garlic, finely chopped; 40g spring onions, segmented; 30g celery, segmented; 50g pickles; 3g sesame; 40g Linjiangsi chili bean paste; 30g homemade spice; 8g salt; 2g MSG; 15ml Shaoxing cooking wine; 3,000ml stock; 80ml cooking oil

PREPARATION

1. Get the meat of the carp, slice and cut the bones into segments. Marinate the meat and bones with salt, Shaoxing cooking wine and MSG.

2. Heat the cooking oil in the wok to 100℃. Blend in Linjiangsi chili bean paste, Sichuan pepper, cooked intestine, spring onions, ginger, and garlic and fry. Add the homemade spice and stock and boil. Boil the fish head and bones in the soup, remove after they are well done and transfer to a pot. Put the fleshes in the wok and boil with low heat. Transfer to the pot and add celery, spring onions and garlic.

3. Heat the cooking oil in the wok to 150℃. Fry dried chili pepper, Sichuan pepper and sesame and sprinkle in the pot.

NOTES

Rivers run through Ziyang City and it has a variety of fishes. Fish with pork Intestines is a popular dish in the city as it looks red and bright and tastes spicy and delicate. The fish is tender, and the intestine is chewy. This dish is not expensive and presents local food characteristics.

19 盘龙鳝鱼
RECLINING DRAGON FINLESS EEL

食材配方

活小鳝鱼1 500克　干辣椒节50克
花椒30克　　　食盐5克
临江寺豆瓣30克　胡椒粉1克
料酒15毫升　　　姜米6克
蒜米10克　　　葱丝3克
十三香2克　　　鸡精2克
味精2克　　　　熟芝麻3克
食用油1 500毫升（约耗80毫升）

INGREDIENTS

1,500g small live finless eels; 50g dried chilies, segmented; 30g Sichuan pepper; 5g salt; 30g Linjiangsi chili bean paste; 1g pepper; 15ml Shaoxing cooking wine; 6g ginger, finely chopped; 10g garlic, finely chopped; 3g spring onions, shredded; 2g thirteen-spice powder; 2g chicken essence; 2g MSG; 3g sesames; 1,500ml cooking oil for deep-frying(about 80ml to be consumed)

制作工艺

1. 鳝鱼用清水饲养1~2天，中途换水多次。
2. 锅置火上，放油烧至200℃，放入鳝鱼炸成金黄色、呈盘龙形状后捞出。
3. 锅置火上，放油烧至120℃，放入临江寺豆瓣、干辣椒节、花椒、姜米、蒜米炒香，再放入鳝鱼、食盐、料酒、十三香、胡椒粉、鸡精、味精炒香，起锅装盘，撒上熟芝麻，点缀上葱丝成菜。

评鉴

此菜因鳝鱼成型似盘龙而得名，具有麻辣干香、回味绵长的特点，食用时应先从头颈部撕至腹部去除内脏。

PREPARATION

1. Purge the finless eels in fresh water for 1 or 2 days, change the water for several times.

2. Heat oil in a wok to 200°C, put finless eels to fry until shaped like reclining dragon with golden brown color, then remove.

3. Heat oil in a wok to 120°C, blend Linjiangsi chili bean paste, dried chilies, Sichuan pepper, ginger, garlic to stir-fry until aromatic. Add finless eels, salt, Shaoxing cooking wine, thirteen-spice powder, pepper, chicken essence and MSG to stir-fry to bring out aroma, remove to a serving dish. Sprinkle with sesames and spring onions then serve.

NOTES

The finless eels shape like reclining dragon after being deep fried, hence the dish's name. It has pungent and aromatic taste, long lasting fragrance.

20 仔姜泥鳅
SOFT POND LOACHES WITH TENDER GINGER

食材配方

泥鳅500克　　仔姜300克
小米辣300克　泡辣椒末250克
大蒜150克　　花椒50克
白糖10克　　 醋20毫升
芥菜花100克　葱花100克
味精3克　　　水淀粉50克
鲜汤500毫升　食用油1000毫升（约耗200毫升）

制作工艺

1 仔姜切成细丝；小米辣切成粗丝。

2 锅置火上，放油烧至180℃，放入泥鳅炸至酥香时捞出。

3 锅置火上，放油烧至120℃，先放入泡辣椒末、花椒、大蒜、仔姜丝、小米辣丝炒香，再加入鲜汤、泥鳅、白糖、醋烧至泥鳅软熟入味，最后加入芥菜花、葱花、味精、水淀粉勾芡，收汁亮油后装盘即成。

评鉴

泥鳅的蛋白质含量高、脂肪低，有降脂、降压等作用，号称"水中人参"。此菜以泥鳅制成，色泽红亮，质地软嫩，味道咸鲜香辣。

INGREDIENTS

500g pond loaches; 300g tender ginger; 300g bird's eye chilies; 250g pickled chilies, finely chopped; 150g garlic; 50g Sichuan pepper; 10g sugar; 20ml vinegar; 100g mustard, finely chopped; 100g spring onions, finely chopped; 3g MSG; 50g batter; 500ml stock; 1,000ml cooking oil (about 200ml to be consumed)

PREPARATION

1. Cut the tender ginger into shreds and bird's eye chilies into thick slivers.

2. Heat the oil to 180℃ and deep fry the pond loaches till crispy and aromatic.

3. Heat the oil to 120℃ and add pickled chilies, Sichuan pepper, garlic, tender ginger slivers, bird's eye chilies. Stir till aromatic and then add stock, pond loaches, sugar and vinegar, brown braise all the ingredients till the pond loaches are soft, cooked through and tasty. Add mustard, spring onions, MSG and coat with batter, stir fry until the seasoning sauce is reduced. Then transfer to the serving dish.

NOTES

Pond loaches are rich in protein and low in fat, which is beneficial for lowering fat and blood pressure. It is known as "the ginseng in water". The dish is made with pond loaches, featuring glistening red color, soft mouth feel, salty, savory and hot tastes.

食材配方

鲜虾300克　　　　小葱200克
红薯馍10个　　　　蒜泥200克
食盐5克　　　　　　酱油15毫升
蚝油10毫升　　　　鸡粉15克
水淀粉20克　　　　鲜汤50毫升

制作工艺

1. 鲜虾去头、壳，保留尾部，从背部剖开，去沙线。
2. 将蒜泥、食盐、酱油、蚝油、鸡粉调匀制成蒜蓉酱。
3. 平底铁锅置火上，小葱切成长约8厘米的节后放入垫底，将虾与蒜蓉酱、水淀粉拌匀后摆在小葱上面，四周配上红薯馍，掺入鲜汤，加盖后上桌加热至虾成熟。

评鉴

此菜借鉴了粤菜中的蒜蓉虾制法，再结合当地口味特点，用川式蒜蓉酱调味，配以当地特产红薯制作的馍。虾仁脆嫩，红薯馍软韧化渣，味道咸鲜、蒜香浓郁，以堂烹形式呈现，不仅菜品好吃好看，更能给客人一种体验感、仪式感。

生焗虾配粗粮锅边馍
21 BAKED PRAWNS WITH SWEET POTATO BUNS

INGREDIENTS

300g fresh prawns; 200g spring onions; 10 sweet potato buns; 200g garlic, smashed; 5g salt; 15ml soy sauce; 10ml oyster sauce; 15g chicken essence; 20g batter; 50ml stock

PREPARATION

1. Remove the heads and shells of the prawns and devein.

2. Mix garlic, salt, soy sauce, oyster sauce and chicken essence and make garlic sauce.

3. Cut the spring onions into about 8cm segments and place in the wok. Mix garlic sauce with batter and put the mix on the segmented spring onions and place sweet potato buns around. Add the stock and cap the wok. Transfer on the table and heat till the prawns are cooked through.

NOTES

This dish is developed from steamed prawns with garlic in Cantonese Cuisine. With Sichuan garlic sauce, sweet potato buns and crispy prawns, this dish tastes tender, salty, fresh and has a strong flavor of garlic. As the wok is heated in front of the guests, they will have a good experience of taking part in the cooking of the dish.

22 CRAYFISH IN CHILI SAUCE
香辣小龙虾

食材配方

小龙虾1 000克　水发魔芋300克
洋葱条150克　　黄瓜条150克
干辣椒节40克　　花椒20克
临江寺豆瓣20克　火锅底料40克
泡辣椒末30克　　野山椒末40克
小米辣末60克　　鸡精3克
味精3克　　　　白糖5克
香料粉15克　　　醋10毫升
藤椒油50毫升　　清水500毫升
食用油2 000毫升（约耗120毫升）

制作工艺

1. 小龙虾去头、虾脑、小脚和虾线，在背上片一刀，用淡盐水清洗干净，入160℃的油中炸至成熟后捞出。

2. 锅置火上，放油烧至120℃，放入干辣椒节、花椒、临江寺豆瓣、泡辣椒末、野山椒末、小米辣末、火锅底料炒香，掺入清水烧沸，入小龙虾、水发魔芋、洋葱条、黄瓜条、白糖、鸡精、味精、香料粉、醋、藤椒油烧至入味，收汁浓稠后起锅装盘即成。

评鉴

资阳中和镇盛产小龙虾，是当地村民采用稻虾、藕田混养的生态模式养殖出来的，单只小龙虾重量平均可达60克，肉质紧致、饱满、细嫩。当地人最传统、最喜爱的菜肴就是香辣小龙虾，色泽红亮，麻辣鲜香，虾肉Q弹。此外，还研发出鱼香味、清蒸双味、红糖糍粑等10余个小龙虾菜肴，深受成渝两地消费者的喜爱。如今，中和镇每年都要举办小龙虾美食节，到中和吃小龙虾已成一种时尚。

INGREDIENTS

1,000g crayfish; 300g konjac tofu, water-soaked; 150g onion, shredded; 150g cucumbers, shredded; 40g dried chilies, segmented; 20g Sichuan pepper; 20g Linjiangsi chili bean paste; 40g hotpot soup base; 30g pickled chilies, chopped; 40g bird's eye chilies, chopped; 60g wild chili peppers, chopped; 3g chicken essence; 3g MSG; 5g sugar; 15g spice powder; 10ml vinegar; 50ml green Sichuan pepper oil; 500ml water; 2,000ml cooking oil(about 120ml to be consumed)

PREPARATION

1. Remove the crayfish's heads, feet and devein them, give a cut on those backs, rinse with light salty water, then deep-fry in 160℃ oil till cooked through.

2. Heat oil in a wok until 120℃, add dried chilies, Sichuan pepper, Linjiangsi chili bean paste, pickled chilies, bird's eye chilies, wild chili peppers and hotpot soup base to stir-fry until aromatic, pour in water to boil, stir in the crayfish, konjac tofu, onions, cucumbers, sugar, chicken essence, MSG, spice powder, vinegar, green Sichuan pepper oil to braise until flavored, heat over a high flame to reduce the sauce then transfer to a serving dish.

NOTES

Crayfish farming is an industry in Zhonghe Town of Ziyang city. Local farmers use rice fields and lotus ponds to raise crayfish. The average weight of a crayfish could reach 60g, and they are meatier, tender and chewier. This dish is local people's favorite traditional food, which has alluring reddish color, spicy and numbing taste and chewier meat. Additionally, they have developed over 10 crayfish dishes popular in Sichuan and Chongqing such as fish-flavor crayfish, steamed double-flavor crayfish and crayfish with brown sugar and glutinous rice cake. Nowadays, Zhonghe Town would hold a Crayfish Festival every year. Having a crayfish meal in Zhonghe Town is a fashion now.

23 红糖糍粑小龙虾
CRAYFISH WITH BROWN SUGAR AND GLUTINOUS RICE CAKE

食材配方

小龙虾500克　糍粑250克
红糖100克　　白糖100克
酥花生碎15克　熟芝麻10克
清水200毫升

制作工艺

1 小龙虾去头、虾脑、小脚和虾线；糍粑切成约2厘米见方的丁；红糖切细。

2 锅置火上，下油烧至160℃，放入小龙虾炸熟后捞出，再入糍粑炸成外表酥脆时捞出。

3 锅置火上，入清水、白糖、红糖，用小火熬制起小泡时放入小龙虾、糍粑丁炒匀，装盘后撒上酥花生碎、熟芝麻即成。

评鉴

此菜是雁江中和镇自创的新口味小龙虾菜品，色泽棕红，小龙虾外酥里嫩，花生、芝麻酥脆，味道甜香不腻，特别适合不嗜辣的人群。

INGREDIENTS

500g crayfish; 250 glutinous rice cake; 100g brown sugar; 100g sugar; 15g fried peanuts, finely chopped; 10g roasted sesames; 200ml water

PREPARATION

1. Remove the heads, brains, small legs of the crayfish and devein. Cut the glutinous rice cake into about 2cm dices. Finely slice the brown sugar.

2. Heat the cooking oil in the wok to 160℃. Fry the crayfish and remove. Fry the glutinous rice cubes till crispy.

3. Add water, sugar and brown sugar in the wok, and simmer till there are small bubbles in the sauce. Put the crayfish and glutinous rice cubes in the pot. Sprinkle with peanuts and sesames and serve at once.

NOTES

Crayfish with Brown Sugar and Glutinous Rice Cake is created in Zhonghe Town, and it is brown reddish. The crayfish, peanuts, and sesame are crispy. The dish tastes sweet and fatty but not greasy and is favored by people who are not able to have spicy dishes.

食材配方

仔鸭500克　　仔姜100克
泡姜25克　　泡辣椒25克
小米辣100克　青尖椒150克
花椒20克　　食盐10克
味精2克　　　鸡精2克
料酒20毫升　　醪醋汁25毫升
芝麻油3毫升　辣椒油20毫升
食用油1 000毫升（约耗60毫升）

制作工艺

1. 仔鸭斩成约2厘米见方的丁；仔姜、泡辣椒、小米辣、青尖椒分别切成小丁；泡姜切成小颗粒。

2. 锅置火上，放油烧至180℃，放入鸭丁爆炒至表面无水分，入花椒、泡姜、泡辣椒、小米辣、青尖椒、仔姜炒熟，再入料酒、醪醋汁、食盐、味精、鸡精、芝麻油、辣椒油炒香即成。

评鉴

仔姜爆鸭是资阳地区常见的一种家常菜品，采用特产的紫竹姜制作成菜，色泽鲜艳，口感干香化渣，味道麻辣咸鲜，香味浓郁。

24 仔姜爆鸭
STIR FRIED DUCK WITH TENDER GINGER

INGREDIENTS

500g duck; 100g tender ginger; 25g pickled ginger; 25g pickled chilies; 100g bird's eye chilies; 150g green chilies; 20g Sichuan pepper; 10g salt; 2g MSG; 2g chicken essence; 20ml Shaoxing cooking wine; 25ml fermented glutinous rice juice; 3ml sesame oil; 20ml chili oil; 1,000ml cooking oil (about 60ml to be consumed)

PREPARATION

1. Cut the duck into about 2cm cubes. Dice the tender ginger, pickled chilies, bird's eye chilies and green chilies. Chop the pickled ginger.

2. Heat oil in a wok to 180℃, and stir fry the duck cubes over high heat to dry their surface water. Add Sichuan pepper, pickled ginger, pickled chilies, bird's eye chilies, green chilies and tender ginger, and continue to stir till cooked through. Blend in the Shaoxing cooking wine, fermented glutinous rice juice, salt, MSG, chicken essence, sesame oil and chili oil, stir till aromatic, and transfer to a serving dish.

NOTES

A household dish in Ziyang, Stir Fried Duck with Tender Ginger uses a local specialty called purple bamboo ginger as one of the main ingredients, features spicy mala tastes and bright colors.

INGREDIENTS

300g clams; 20g pickled bird's eye chilies; 10g garlic, finely chopped; 50g Chinese leeks; 10g green garlic; 2g cumin powder; 6g salt; 6ml Shaoxing cooking wine; 20g batter; 3ml sesame oil; 1,000ml cooking oil (about 60ml to be consumed).

食材配方

河蚌300克	泡小米辣20克
蒜粒10克	韭菜50克
蒜苗10克	孜然粉2克
食盐6克	料酒6毫升
水淀粉20克	芝麻油3毫升
食用油1 000毫升（约耗60毫升）	

制作工艺

1. 河蚌取肉、切片；泡小米辣切小颗粒；韭菜、蒜苗分别切碎。
2. 河蚌肉加食盐、料酒、水淀粉码味上浆，入锅炸熟后捞出。
3. 将食盐、孜然粉、料酒、水淀粉入碗调成芡汁。
4. 锅置火上，放油烧至150℃，放入泡小米辣、蒜粒、蒜苗、韭菜炒香，再入河蚌炒匀，倒入芡汁，收汁亮油后装盘即成。

评鉴

资阳境内河流、池塘、稻田众多，盛产河蚌，民间多用它入菜，不但丰富了餐桌，还兼具清热解毒、明目的作用。此菜以河蚌为食材制成，味道咸鲜微辣。

PREPARATION

1. Slice the meat of the clams. Chop the chilies, Chinese leeks and green garlic.

2. Roll the clam meat in salt, Shaoxing cooking wine and batter, fry and remove from the wok.

3. Mix salt, cumin powder, Shaoxing cooking oil and batter in a bowl to make the thickening sauce.

4. Heat the oil in the wok to 150°C. Fry chilies, garlic, green garlic, and Chinese leeks, and add the clams and thickening sauce, simmer till the juice is reduced. Then serve the dish.

NOTES

There are many rivers, ponds and fields in Ziyang and abundant clams are growing there. Locals use clams for meals, which is good for people's health. The dish tastes salty and a bit spicy.

DRY 25 干钠河蚌
BRAISED CLAMS

26 桑茶汁鸡豆花

CHICKEN DOUFU IN MULBERRY TEA

食材配方

鸡脯肉250克　　鸡蛋清200克
水淀粉100克　　鸡粉5克
味精2克　　　　胡椒粉0.5克
食盐5克　　　　料酒10毫升
特制清汤500毫升　桑叶茶水300毫升
桑叶茶芽3克　　红鱼子酱10克
清水300毫升　　姜葱水30毫升

制作工艺

1. 鸡脯肉加工成泥状，去筋，加入姜葱水、鸡蛋清、食盐、料酒、胡椒粉、水淀粉、鸡粉、味精、清水搅匀成鸡浆。

2. 锅置火上，掺入特制清汤烧沸，倒入鸡浆，用小火慢煮至成熟，待呈豆花状后舀入盛有桑叶茶水的茶碗中，灌入清汤，点缀上红鱼子酱、桑叶茶芽成菜。

评鉴

鸡豆花是川菜最具代表性的经典菜品之一，常在高级别的宴会和国宴中出现，具有汤清菜白、咸鲜味醇、质地细嫩、形似豆花的特点。此菜在传统鸡豆花的基础上加入当地人工种植的可食用桑叶茶水和桑叶茶芽，色碧绿清香，抗氧化能力强，别有风味。

INGREDIENTS

250g chicken breast; 200g egg white; 100g batter; 5g chicken essence; 2g MSG; 0.5g pepper; 5g salt; 10ml Shaoxing cooking wine; 500ml homemade consommé; 300ml mulberry tea; 3g mulberry leaf sprouts; 10g red caviar; 300ml water; 30ml ginger-and-scallion-flavored juice

PREPARATION

1. Mince the chicken breast and dislodge the tendon, add the ginger-and-scallion-flavored juice, egg white, salt, Shaoxing cooking wine, pepper, batter, chicken essence, MSG and water, then mix them well.

2. Heat the pot and add homemade consommé. Bring to a boil, add all the mixed ingredients and stew them over a low flame till cooked through like the shape of doufu. Transfer to the tea bowl containing mulberry leaf tea and add consommé. Garnish with red caviar and mulberry leaf sprouts.

NOTES

Chicken Doufu is one of the most classic soup in Sichuan cuisine, which usually appears in the high-level banquets and state banquets. The soup is limpid and the food materials are light in color. It is salty, delicate and tasty in flavor, tender and soft in mouth feel and looks like the shape of doufu. The dish adds mulberry tea and mulberry leaf sprouts on the basis of traditional chicken doufu, which is light in color and has a strong aroma of tea.

27 鸡蒙竹荪
BAMBOO FUNGUS COATED WITH CHICKEN

食材配方

鸡脯肉150克　　竹荪20克
番茄片75克　　　菜心50克
鸡蛋清120克　　水淀粉50克
味精2克　　　　胡椒粉0.5克
食盐5克　　　　料酒10毫升
清汤1 000毫升　姜葱水100毫升
食盐3克　　　　化猪油20毫升

INGREDIENTS

150g chicken breast; 20g veiled bamboo fungus; 75g tomatoes, sliced; 50g choy sum; 120g egg white; 50g batter; 2g MSG; 0.5g pepper; 5g salt; 10ml Shaoxing cooking wine; 1,000ml consommé; 100ml ginger-and-scallion flavored juice; 3g salt; 20ml lard

制作工艺

1 鸡脯肉加工成泥状，去筋，加入姜葱水、鸡蛋清、食盐、料酒、胡椒粉、水淀粉、味精、化猪油搅拌成鸡糁。

2 竹荪切成长约6厘米的段，将鸡糁平铺在竹荪上，入沸水锅中煮熟后捞出，盛入碗中，灌上烧沸的清汤，加入焯水后的菜心、番茄片即成。

PREPARATION

1. Mince the chicken breast and dislodge the tendon, add ginger-and-scallion flavored juice, egg white, salt, Shaoxing cooking wine, pepper, batter, MSG and lard, then mix them well.

2. Cut the bamboo fungus into about 6cm segments and put the mixed ingredients on the bamboo fungus segments. Put in the boiling water till they are cooked through. Transfer to the serving bowl and pour the boiling consommé over. Then add the parboiled choy sum and tomatoes slices.

评鉴

此菜是在鸡蒙菜心的基础上演变而来，色泽自然，汤清澈见底，味咸鲜醇厚。

NOTES

The dish is developed on the basis of Choy Sum Coated with Chicken. It is natural in color and lustre with limpid soup and salty-savory flavor.

28 柠檬鸡豆花

LEMON FLAVORED CHICKEN DOUFU

食材配方

鸡脯肉300克
菜心10克
特制清汤1 000毫升
料酒10毫升
味精3克
柠檬汁15毫升
鸡蛋清150克
食盐20克
姜葱水20毫升
水淀粉150克

制作工艺

1. 鸡脯肉加工成蓉，加水、柠檬汁、食盐、料酒、姜葱水、味精、水淀粉、鸡蛋清，搅拌成鸡浆。

2. 锅置火上，倒入特制清汤烧沸，用炒勺搅动清汤，迅速倒入鸡浆，用小火燉制30分钟左右，至鸡浆凝结成团、成熟后装入汤碗，放入烫熟的菜心即成。

评鉴

此菜在川菜传统名菜鸡豆花的基础上加入柠檬汁，形如豆花，口感软滑细嫩，色泽更加洁白，味道更加咸鲜清爽。

INGREDIENTS

300g chicken breast; 15ml lemon juice; 10g choy sums; 150g egg white; 1,000ml homemade consommé; 20g salt; 10ml Shaoxing cooking wine; 20ml ginger-and-scallion-flavored juice; 150g batter

PREPARATION

1. Mince the chicken breast. Add water, lemon juice, salt, Shaoxing cooking wine, ginger and ginger-and-scallion-flavored juice, MSG, batter, and egg white, and mix with the chicken paste.

2. Pour the homemade consommé in the wok and boil. Stir and blend in the chicken paste. Simmer for 30 minutes until the chicken paste is condensed into masses and transfer into a soup bowl. Serve at once with the cooked choy sums.

NOTES

Lemon Flavored Chicken Doufu is developed from Chicken Doufu, a famous Sichuan dish and lemon juice is added to the latter. Lemon Flavored Chicken Doufu is like traditional doufu and it is white, tastes soft, smooth, salty and delicate.

29 贵凤穿柠衣
PHOENIX COATED IN LEMON

食材配方

鸡腿肉250克　柠檬200克　姜5克
蒜5克　小米辣椒10克　青尖椒20克
食盐5克　料酒5毫升　白糖3克
味精2克　水淀粉20克　食用油40毫升
葱20克

制作工艺

1 鸡腿肉切成丁，加入食盐、料酒、水淀粉拌匀；柠檬取皮，切成菱形块；姜、蒜切成指甲片；葱、小米辣椒、青尖椒切成马耳朵形。

2 将食盐、料酒、白糖、味精、水淀粉入碗调成芡汁。

3 锅置火上，放油烧至180℃，先放入鸡丁炒熟，再放入柠檬皮块、姜、葱、小米辣椒、青尖椒炒香，倒入芡汁，收汁亮油后起锅装盘即成。

评鉴

此菜是在贵州鸡制作的基础上改进而成。鸡俗称"凤"，与柠檬皮一起烹制，如同穿了一件外衣，故此得名。此菜色泽淡雅，肉质细嫩，味咸鲜略酸，且有柠檬特殊的清香味。

INGREDIENTS

250g boneless chicken leg; 200g lemon; 5g ginger; 5g garlic; 10g bird's eye chilies; 20g green chilies; 5g salt; 5ml Shaoxing cooking wine; 3g sugar; 2g MSG; 20g batter; 40ml cooking oil; 20g spring onions

PREPARATION

1. Dice the chicken, and mix well with salt, Shaoxing cooking wine and batter. Peel the lemons to get the rind, and cut into diamond cubes. Cut the ginger and garlic into thumbnail slices, and the spring onions, bird's eye chilies and green chilies into horse-ear slices.

2. Combine salt, Shaoxing cooking wine, sugar, MSG and batter in a bowl to make the seasoning sauce.

3. Heat oil in a wok to 180°C, and stir fry the chicken, lemon rind, ginger, spring onions, bird's eye chilies and green chilies till aromatic. Pour in the seasoning sauce, and braise to reduce the sauce till the oil is crystal clear. Transfer to a serving dish.

NOTES

A modified version of Guizhou Chicken, the dish is so named because chicken (traditionally called phoenix), when cooked with lemon rind, looks as if it is wearing a lemon coat. This appealing dish is not only pleasant to the eye, but also mouthwatering with its tender, salty, slightly sour chicken highlighted with lemon fragrance.

30 无名烧鸡 WUMING BRAISED CHICKEN

食材配方

土鸡2 000克	干辣椒段8克
花椒2克	姜片15克
仔姜片50克	葱段20克
青尖椒段500克	蒜片15克
食盐3克	白糖3克
味精2克	熟芝麻5克
料酒10毫升	鲜汤2 000毫升
食用油80毫升	

制作工艺

1. 将鸡肉斩成约3厘米见方的块，加入食盐、姜片、葱段、料酒拌匀码味。

2. 锅置火上，放入油烧至80℃，放入干辣椒段、花椒炒香，再入鸡块、姜片炒香，掺入鲜汤烧沸，入高压锅压5分钟。

3. 锅置火上，放入油烧至120℃，放入青尖椒段、蒜片、仔姜片炒香，加入鸡块及汁水，用中火收汁浓稠后，放入白糖、味精调味，起锅装盘后撒上熟芝麻即成。

评鉴

此菜以农家散养的土鸡为食材，采用红烧的烹调方法制作而成，色泽红亮鲜艳，味道鲜香麻辣，口感软糯。

INGREDIENTS

2,000g free-range chicken; 8g dried chilies, segmented; 2g Sichuan pepper; 15g ginger, sliced; 50g tender ginger; 20g spring onions, segmented; 500g green chilies, segmented; 15g garlic, sliced; 3g salt; 3g sugar; 2g MSG; 5g roasted sesames; 10ml Shaoxing cooking wine; 2,000ml stock; 80ml cooking oil

PREPARATION

1. Cut the chicken into chunks of about 3cm long. Season with salt, ginger, spring onions and Shaoxing cooking wine.

2. Heat the oil in a wok to 80°C and stir fry dried chilies and Sichuan pepper till aromatic. Add chicken chunks and ginger to fry till aromatic, add stock, bring to a boil and cook in the pressure cooker for 5 minutes.

3. Heat the oil in a wok to 120°C, stir fry green chilies, garlic and tender ginger till aromatic. Add chicken chunks and chicken broth, braise over a medium flame till the sauce is reduced. Then add sugar and MSG and transfer to the serving dish, sprinkle roasted sesames.

NOTES

This dish uses free-range chicken raised by farmers as the basic ingredient and is cooked by brown braising. Its color is bright red and it is numbing and spicy, which tastes soft and sticky.

31 仔姜烧老鸭
BRAISED DUCK WITH TENDER GINGER

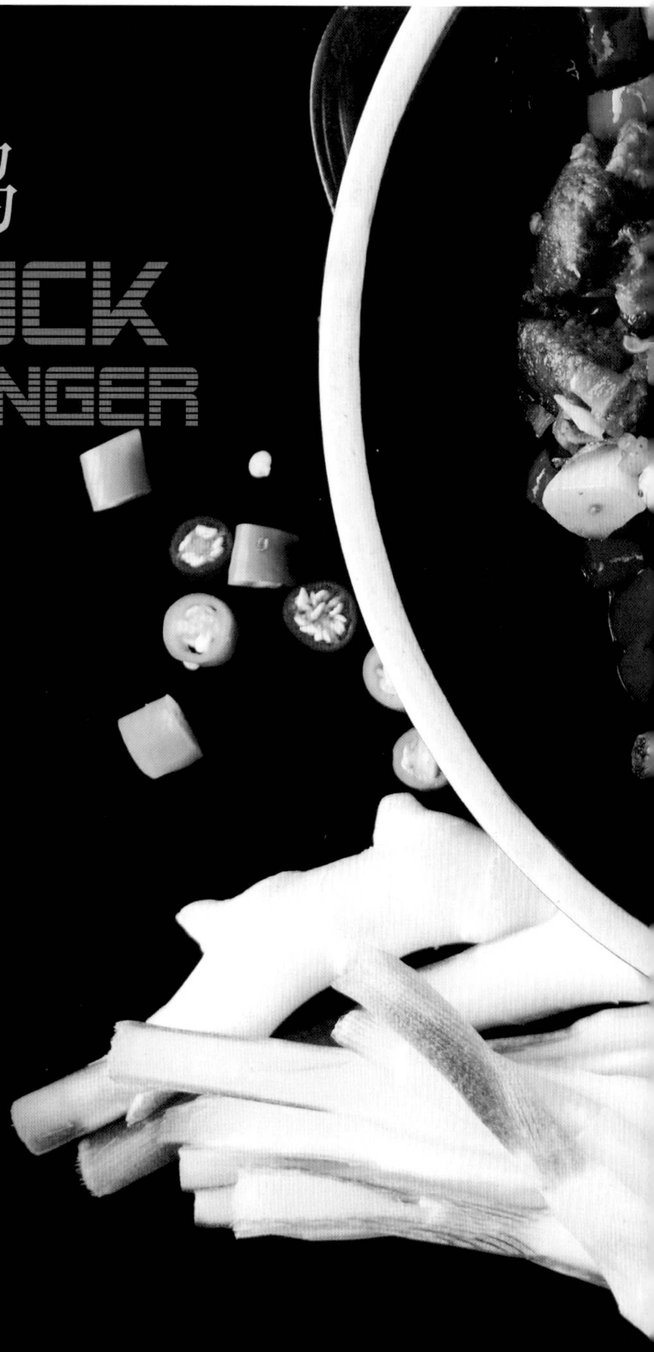

食材配方

老鸭1000克　　泡姜40克
大蒜30克　　　仔姜100克
青尖椒30克　　葱段10克
花椒5克　　　　干辣椒节10克
临江寺豆瓣50克　白糖3克
蚝油3毫升　　　料酒20毫升
八角3克　　　　食盐3克
味精2克　　　　水淀粉20克
鲜汤1000毫升　　食用油150毫升

制作工艺

1 将老鸭斩成约3厘米见方的块，用清水浸泡去除血污后捞出，沥干水分。

2 仔姜、泡姜切成厚约0.3厘米的片；青尖椒切成长约4厘米的段。

3 锅置火上，放油烧至180℃，放入老鸭煸炒出香，放入泡姜、干辣椒节、花椒、八角、临江寺豆瓣炒香，掺入鲜汤，放入食盐、料酒、白糖烧至鸭肉软熟入味，再放入青尖椒段、大蒜、葱段、蚝油、仔姜烧熟，最后入味精，用水淀粉收汁，起锅装盘即成。

评鉴

此菜以安岳县鼎新乡出产的仔姜为辅料，与老鸭同烧成菜，色泽红亮，鸭肉质地软糯，味道咸鲜麻辣、仔姜的辛香味突出。

INGREDIENTS

1,000g duck; 40g pickled ginger; 30g garlic; 100g tender ginger; 30g green chilies; 10g spring onions, segmented; 5g Sichuan pepper; 10g dried chilies, segmented; 50g Linjiangsi chili bean paste; 3g sugar; 3g oyster sauce; 20ml Shaoxing cooking wine; 3g star anise; 3g salt; 2g MSG; 20g batter; 1,000ml stock; 150ml cooking oil

PREPARATION

1. Cut the duck into about 3cm chunks and soak in water. Then remove and drain the water.

2. Cut the tender ginger and pickled ginger into about 0.3cm slices, and cut the green chilies into about 4cm segments.

3. Heat the oil to 180°C in a wok, and add the duck and stir fry till aromatic, then add pickled ginger, dried chilies, Sichuan pepper, star anise and Linjiangsi chili bean paste, continue to stir. Add the stock, salt, Shaoxing cooking wine and sugar, stew till the duck is cooked through. Add green chilies segments, garlic, spring onions, oyster sauce and tender ginger, stew till cooked through. Add MSG and batter, stir fry till the sauce is reduced. Then transfer to the serving dish.

NOTES

This dish is made from tender ginger, freshly produced in Dingxin Township, Anyue county, and stir fried with duck. Its color is bright red and it tastes soft, sticky, salty, savory and spicy with a strong pungent flavor of the tender ginger.

32 STEAMED SQUABS IN PEA STOCK
养生豆汤乳鸽

食材配方

乳鸽500克　　莲子20克
山药20克　　　红腰豆20克
芦笋20克　　　玉米20克
食盐10克　　　味精2克
豆汤750毫升

制作工艺

1. 乳鸽斩成约4厘米见方的块，焯水后备用；山药、芦笋切成丁。
2. 将乳鸽放入大蒸碗中，加入豆汤、莲子，入笼蒸至乳鸽软熟，再放入山药、红腰豆、芦笋、玉米、食盐、味精，蒸10分钟成菜。

评鉴

豆汤饭是四川人冬天最喜爱的美食之一。鸽子肉属于高蛋白、低脂肪，有养血补气等作用。此菜将乳鸽用豆汤蒸制成菜，味道咸鲜，具有养生之功效。

INGREDIENTS

500g squabs; 20g lotus seeds; 20g Chinese yams; 20g red kidney beans; 20g asparagus; 20g corn; 10g salt; 2g MSG; 750ml pea stock

PREPARATION

1. Cut the squabs into about 4cm square cubes, blanch. Dice the Chinese yams, asparagus.
2. Put the squabs into a large steam bowl, add pea stock and lotus seeds to steam until the meat is soft, then add Chinese yams, red kidney bean, asparagus, corn, salt, MSG, then steam another 10 minutes.

NOTES

Rice in pea stock is one of Sichuanese favorite foods in winter. Pigeon meat has high protein but low fat, which has great benefits for our health. This dish uses squabs steamed in pea stock, combining the taste and health benefits perfectly.

33 苦荞鸭汤锅
DUCK SOUP WITH KUJIAO

食材配方

土鸭750克　　苦荞250克
猪肚250克　　姜片5克
葱段3克　　　西红柿5克
红枣5克　　　枸杞2克
食盐12克　　　味精3克
猪骨汤2 000毫升

制作工艺

1. 猪肚焯水后切成条；土鸭斩成块，焯水后备用。

2. 砂锅置火上，放入鸭块、肚条、姜片、葱段、红枣、枸杞、猪骨汤、苦荞，先用旺火烧沸，再改用小火煨至鸭肉软熟，最后入西红柿、食盐、味精调味即成。

评鉴

苦荞能清热、健胃，鸭肉能消热止渴、补虚益脾，两者结合是夏天滋补汤锅的首选。此菜鸭肉软熟、苦荞微苦，汤清淡爽口，深得百姓认可和喜爱。

INGREDIENTS

750g free-range duck; 250g kujiao bulbs; 250g pork tripe; 5g ginger, sliced; 3g spring onions, segmented; 5g tomatoes; 5g dates; 2g wolfberries; 12g salt; 3g MSG; 2,000ml pork bone stock

PREPARATION

1. Blanch the pork tripe and slice into strips. Cut the duck into chunks and blanch.

2. Blend in a pot the duck chunks, tripe strips, ginger, spring onions, dates, wolfberries, pork bone stock and kujiao. Bring to a boil over high heat, reduce to low heat and continue to simmer till the duck is soft. Add tomatoes, salt and MSG to the soup.

NOTES

Kujiao, a daily herb, can help clear the heat in body and invigorate the stomach. The duck can help restore the deficiency of the body and invigorate the spleen. It is the first choice of locals in summer. In this dish, the duck is soft, kujiao is a little bitter and the soup is fresh and favored by the masses.

食材配方

猪瘦肉400克	鲜柠檬皮50克
姜丝30克	干辣椒丝20克
食盐5克	临江寺豆瓣30克
花椒粉1克	辣椒粉5克
芝麻5克	味精2克
料酒15毫升	食用油500毫升（约耗60毫升）

制作工艺

1 鲜柠檬皮切成二粗丝；猪瘦肉切成粗丝。

2 锅置火上，放油烧至180℃，入猪肉丝炸至表面酥香、色浅黄时捞出；锅中留少量余油，放入肉丝、姜丝、柠檬丝、料酒煸炒出香味，入临江寺豆瓣炒香油呈红色，再入辣椒粉、干辣椒丝、芝麻、食盐、花椒粉、味精炒香，出锅装盘即成。

评鉴

此菜将鲜柠檬皮作为干煸肉丝的辅料，肉丝干香软绵，色彩更加鲜艳，味道香辣浓厚，柠檬香味突出。

INGREDIENTS

400g lean pork; 50g fresh lemon skin; 30g sliced ginger; 20g dried chilies, sliced; 5g salt; 30g Linjiangsi chili bean paste; 1g ground Sichuan pepper; 5g ground chilies; 5g sesames; 2g MSG; 15ml Shaoxing cooking wine; 500ml cooking oil for deep-frying (about 60ml to be consumed)

PREPARATION

1. Cut the fresh lemon skin into medium slivers, cut the lean pork into thick slivers.

2. Heat oil in a wok to 180℃, add pork to fry until the surface is crispy with light golden color, then remove from the heat. Keep small amount of oil in the wok, add the pork, ginger, lemon skin, Shaoxing cooking wine and stir-fry to bring out the aroma, blend with Linjiangsi chili bean paste to fry until aromatic and the oil becomes reddish brown, add ground chilies, dried chili, sesames, salt, ground Sichuan pepper, MSG to stir-fry until aromatic, remove from the heat and transfer to a serving dish.

NOTES

This dish not only has tender pork slivers, rich savory and spicy taste, but also brighter color and strong lemon aroma by using fresh lemon skin in it.

34 柠皮煸肉丝
STIR-FRIED PORK WITH LEMON RIND

35 旱蒸坛子肉
STEAMED TANZI PORK

食材配方

坛子肉250克

水发干豇豆200克

香菜1根

制作工艺

1 坛子肉切成长约7厘米、宽约4厘米、厚约0.3厘米的片；水发干豇豆切成长约4厘米的段。

2 将干豇豆段入笼垫底，坛子肉片摆放在面上成三叠水形，入锅蒸制10分钟，出笼后，点缀上香菜即成。

评鉴

坛子肉是资阳市著名特产，境内各区县都有制作，极富地方特色。坛子肉营养丰富、风味独具一格，广泛用于菜点的制作，可蒸、炒、焖、烧。此菜采用旱蒸的烹饪方法制成，荤素搭配合理，色彩分明，肉质粑糯，味道咸鲜宜人。

INGREDIENTS

250g Tanzi Pork; 200g water soaked dry long beans; 1 stalk of coriander

PREPARATION

1. Cut the Tanzi Pork into about 7cm-long, 4cm-wide and 0.3cm-thick slices, and long beans into about 4cm sections.

2. Place the segmented long bean sections on the bottom of the steamer, and stack on them the pork slices in three layers. Steam for 10 minutes. Remove from the steamer, and garnish with coriander.

NOTE

Tanzi Pork, a common local specialty of Ziyang, is widely used as an ingredient with its rich nutrition and peculiar flavors. It can be steamed, stir fried, pressure simmered and braised. This dish features bright colors, balanced nutrition, and soft, salty, delicate pork.

36 回锅坛子肉
TWICE-COOKED TANZI PORK

食材配方

坛子肉300克　　蒜苗50克
青尖椒100克　　红尖椒20克
味精1克　　　　食用油30毫升

制作工艺

1. 坛子肉入笼蒸熟，晾凉后切成厚约0.3厘米的片；青尖椒、红尖椒、蒜苗切成马耳朵形。
2. 锅置火上，放油烧至160℃，放入坛子肉片炒香，再放入青尖椒、红尖椒、蒜苗炒熟出香，入味精炒匀，起锅装盘即成。

评鉴

此菜采用传统回锅肉的烹调方法烹制而成，肉味咸鲜醇香，肥而不腻，风味独特。

INGREDIENTS

300g Tanzi Pork; 50g green garlic; 100g green chilies; 20g red chilies; 1g MSG; 30ml cooking oil

PREPARATION

1. Steam the pork in the steamer and slice into about 0.3cm thick pieces after cooling. Cut the green chilies, red chilies and green garlic like horse ears.

2. Heat the cooking oil in the wok to 160°C. Stir fry the pork until aromatic. Blend in green chilies, red chilies and green garlic and fry. Add MSG, blend well, and transfer to a serving dish.

NOTES

This dish is prepared with the method of twice cooking. It tastes salty, mellow, fatty but not greasy, and is favored by the masses.

37 脆炸酥肉
CRISPY DEEP FRIED PORK

食材配方

五花肉200克　　红薯粉150克
鸡蛋2个　　　　刀口花椒2克
姜米8克　　　　辣椒粉5克
食盐5克　　　　老抽3毫升
椒盐10克　　　　菜籽油1 500毫升

制作工艺

1. 将五花肉切成长约6厘米、宽约4厘米、厚约0.5厘米的片，放入食盐、老抽、刀口花椒、姜米、辣椒粉、红薯粉搅匀，再放入鸡蛋，加沸水后搅拌均匀。

2. 锅置火上，放入菜籽油烧至180℃，放入肉片炸至外酥脆、内细嫩后捞出装盘，配上椒盐味碟成菜。

评鉴

此菜选用安岳特产的红薯粉制作而成，色泽金黄，外酥内嫩，咸鲜麻辣。

INGREDIENTS

200g pork belly; 150g sweet potato starch; 2 eggs; 2g Sichuan pepper, roasted and chopped; 8g ginger, finely chopped; 5g ground chilies; 5g salt; 3ml dark soy sauce; 3g pepper salt; 1,500ml rapeseed oil

PREPARATION

1. Cut the pork belly into about 6cm-long, 4cm-wide and 0.5cm-thick slices. Mix well with salt, soy sauce, Sichuan pepper, ginger, chilies and sweet potato starch. Add beaten egg and boiling water, and stir well.

2. Heat the rapeseed oil in a wok to 180°C, and deep fry the pork slices till tender on the inside and crispy on the outside. Transfer to a serving dish, and serve with the pepper salt.

NOTES

Anyue sweet potato starch, a local specialty, is used as an ingredient. The dish features golden brown color and a salty and spicy taste. The pork is crispy on the surface, but soft and tender on the inside.

38 乐至烤肉 LEZHI PORK BBQ

食材配方

五花肉500克　食盐5克
姜末2克　鸡精2克
料酒5毫升　自制香料粉5克
食用油50毫升

制作工艺

1 将五花肉切成长约10厘米、宽约1.5厘米的片，加入食盐、姜末、鸡精、料酒、自制香料粉、食用油拌匀，码味4小时，用竹签穿上。

2 将环保碳放至烤炉中烧至过心，再将码好味的五花肉串放至烤炉上反复烤为金黄色即成。

评鉴

乐至烧烤以其香、辣、麻、爽口不腻等特色获得乐至县十大特产称号，深受大众喜爱，其制作技艺已列入资阳市第四批非物质文化遗产代表性名目。乐至烧烤的经典品种是烤五花肉，此外还可以烤制脆骨、牛肉、羊肉等众多食材。

INGREDIENTS

500g pork belly; 5g salt; 2g ginger, chopped; 2g chicken essence; 5ml Shaoxing cooking wine; 5g homemade spices powder; 50ml cooking oil

PREPARATION

1. Cut the pork belly into about 10cm long and 1.5cm wide slices, add salt, chopped ginger, chicken essence, Shaoxing cooking wine, homemade spices powder and cooking oil and blend well, marinate for 4 hours, then skewer the pork slices with bamboo sticks.

2. Light the barbecue charcoals in a grill until it burns red, and grill the pork belly kebabs and turn around several times till it is golden brown, then remove and serve.

NOTES

With its aromatic, spicy, numbing, fatty but not greasy features, Lezhi BBQ ranks among the top 10 local specialties, which is widely popular with the public. Its cooking skills have been listed in the 4th batch of Intangible Heritage Representative List of Ziyang City. Pork belly is the classic dish of Lezhi BBQ, however, the ingredients include but not limit to gristle, beef, lamb, and so on.

39 绿毛竹烤肉
GRILLED PORK IN GREEN BAMBOO TUBES

食材配方

猪碎肉500克　食盐5克
姜末2克　　　胡椒粉1克
自制料粉5克

制作工艺

1. 猪碎肉加工成末，加入食盐、姜末、胡椒粉、自制料粉拌匀成肉馅。
2. 取一端留有竹节的新鲜绿毛竹洗净，将肉馅灌入，放入烤炉上烤熟即成。

评鉴

此菜的自制料粉选用藿香、马蹄草等磨细而成，有开胃健脾、消食化积等作用，与猪肉一起放入绿毛竹中烤制，味道咸鲜，香味浓郁。

INGREDIENTS

500g pork mince; 5g salt; 2g ginger, chopped; 1g pepper; 5g homemade spices powder

PREPARATION

1. Mix pork mince with salt, ginger, pepper and self-made spices powder into the filling.

2. Rinse fresh green bamboo stems with a node on one side, stuff with the filling, and roast in an oven until cooked through.

NOTES

This dish's homemade spices powder is made from huoxiang herb and water shield plants, which has benefits for our appetites and digestions. The green bamboo stems give the pork mince fresh bamboo smell, and this dish has salty and savory taste.

宫保坨子肉
40 GONGBAO PORK BELLY

食材配方

五花肉250克　酥腰果100克
熟芝麻5克　　八角3克
山柰3克　　　糖色50克
干辣椒30克　　花椒2克
姜片15克　　　蒜片20克
葱丁40克　　　食盐2克
白糖10克　　　醋8毫升
老抽2毫升　　料酒6毫升
生粉50克　　　水淀粉30克
鲜汤25毫升　　食用油100毫升
清水500毫升

制作工艺

1. 五花肉切成长约4厘米、粗约2厘米见方的块。
2. 将食盐、白糖、老抽、醋、料酒、水淀粉、鲜汤调成芡汁。
3. 锅置火上，放油烧至160℃，放入五花肉、八角、山柰、姜片、葱丁煸炒至出香吐油，加入糖色、清水烧至软糯时出锅。
4. 锅置火上，放油烧至180℃，将肉块与生粉拌匀后放入，炸至外酥内嫩时捞出。
5. 锅置火上，放油烧至130℃，放入干辣椒、花椒、姜片、蒜片、葱丁炒香，再入肉块炒匀，倒入芡汁，入酥腰果、熟芝麻略炒，起锅装盘即成。

评鉴

此菜在传统宫保菜肴的基础上演变而来，色泽棕红，味道咸鲜麻辣，略带甜酸，肉丁外酥内嫩，花生香脆。

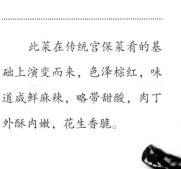

INGREDIENTS

250g pork belly; 100g cashew nuts; 5g roasted sesames; 3g star anise; 3g sand ginger; 50g caramel; 30g dried chili peppers; 2g Sichuan pepper; 15g ginger, sliced; 20g garlic, sliced; 40g spring onion, chopped; 2g salt; 10g sugar; 8ml vinegar; 2ml dark soy sauce; 6ml Shaoxing cooking wine; 50g cornstarch; 30g batter; 25ml stock; 100ml cooking oil; 500ml water

PREPARATION

1. Slice the pork into strips about 4cm long and 2cm thick.

2. Make thickening sauce out of salt, sugar, dark soy sauce, vinegar, Shaoxing cooking wine, batter and stock.

3. Place the wok on the stove and heat the cooking oil in it to 160℃. Blend in pork belly, anis, ginger, sand ginger, and spring onions and fry. Add caramel and water and braise until soft.

4. Place the wok on the stove and heat the cooking oil in it to 180℃. Mix the meat with cornstarch and deep fry till crispy.

5. Place the wok on the stove and heat the cooking oil in it to 130℃. Stir fry dried chili pepper, Sichuan pepper, ginger, garlic, and spring onions. Blend in the pork dices and thickening sauce. Add cashew nuts and sesames, stir briefly and transfer to a serving dish.

NOTES

Gongbao Pork Belly is developed from traditional Gongbao dishes. It is reddish brown and tastes salty, fresh, spicy, and a bit sweet. The meat is tender and the peanuts are crispy.

WULIANG BROWN BRAISED PORK
WITH FERMENTED GLUTINOUS RICE

食材配方

五花肉500克　西瓜300克
醪糟50克　　红曲米10克
葱段35克　　姜片20克
香菜头15克　冰糖30克
糖色30克　　食盐5克
生抽8毫升　　花雕酒30毫升
蚝油30毫升　香醋3毫升
清水600毫升　食用油2 000毫升（约耗30毫升）

制作工艺

1 将五花肉煮至断生，切成约4厘米见方的块，放入200℃油中炸至金黄色时捞出；西瓜挖成小球；姜片、葱段、香菜头放入150℃油中炸香后捞出。

2 锅置火上，加入清水、冰糖、姜片、葱段、香菜头、醪糟、花雕酒、红曲米、糖色、食盐、生抽、蚝油、五花肉，用中火烧15分钟，再改用小火煨制75分钟，捞出五花肉。

3 锅置火上，放入去渣后的汤汁，入五花肉，用小火收汁亮色，滴入香醋，装入盘中，放入西瓜球即成。

评鉴

"无量醪糟红烧肉"因著名书法家谢无量1909年回家乡乐至探亲，在回澜镇陈姓绅士府中食用后赞不绝口而得名。此菜色泽红亮，质地软糯，味道咸甜，肥而不腻。

INGREDIENTS

500g pork belly; 300g watermelon; 50g fermented glutinous rice; 10g red yeast rice; 35g spring onions, segmented; 20g ginger, sliced; 15g heads of coriander; 30g rock sugar; 30g caramel; 5g salt; 8ml light soy sauce; 30ml Shaoxing Hua Tiao Chiew; 30ml oyster sauce; 3ml vinegar; 600ml water; 2,000ml cooking oil (about 30ml to be consumed)

PREPARATION

1. Boil the pork belly until it is medium-well, then cut into 4cm small chunks. Heat the oil to 200°C and deep-fry the pork chunks till golden. Scoop out the watermelon to be in little ball shape. Deep-fry ginger, spring onion and heads of green coriander in oil heated to 150°C till aromatic.

2. Heat the wok and add water, rock sugar, ginger, spring onions, heads of green coriander, fermented glutinous rice, Shaoxing Hua Tiao Chiew, red yeast rice, caramel, salt, light soy sauce, oyster sauce and pork belly. Boil them over a medium flame for 15 minutes, then simmer for 75 minutes and remove the pork belly.

3. Heat the wok and add the residue-free soup and the pork belly. Simmer over a low flame till the sauce is reduced, then add vinegar and the watermelon balls.

NOTES

Wuliang Brown Braised Pork with Fermented Glutinous Rice became well-known because it was highly praised by a famous calligrapher Xie Wuliang. In 1,909, when Xie Wuliang returned to his hometown Lezhi County to visit his relatives, he enjoyed the dish in the mansion of a local squire in Huilan Town. This dish is bright in color and soft in quality, and it tastes salty and sweet, meanwhile, it is fatty but not greasy.

42 酸鲊肉
SUANZHA PORK

食材配方

带皮五花肉250克　熟米粉150克
食盐5克　　　　　料酒10毫升
甜面酱3克　　　　白糖3克
味精2克　　　　　花椒粉2克
辣椒粉3克　　　　菜籽油20毫升

制作工艺

1 将带皮五花肉切成厚约0.3厘米的片，加入食盐、料酒、甜面酱、白糖、味精、花椒粉、辣椒粉、熟米粉、菜籽油拌匀，入坛密封腌制20天后取出。

2 将腌制后的肉摆放在蒸碗中呈一封书形，入笼用旺火蒸熟后取出，翻扣在盘中成菜。

评鉴

此菜是安岳县农村常见做酸鲊肉的一种方法，风味的生成主要取决于肉入坛中腌制的时间和温度，春秋季节一般为20天左右，夏季和冬季可以适当缩短或延长。成菜具有色泽红亮、香味浓郁、酸香适口、肥而不腻的特点。用此法还可以制作酸鲊鱼、酸鲊牛肉等菜品。

INGREDIENTS

250g pork belly with skin attached; 150g cooked ground rice; 5g salt; 10ml Shaoxing cooking wine; 3g fermented flour paste; 3g sugar; 2g MSG; 2g ground Sichuan pepper; 3g ground chilies; 20ml rapeseeds oil

PREPARATION

1. Cut the pork into about 0.3cm thick slices. Add salt, Shaoxing cooking wine, fermented flour paste, sugar, MSG, ground Sichuan pepper, ground chilies, cooked ground rice and rapeseeds oil and mix well. Seal the pork in a pottery pickle jar to marinate for 20 days then remove.

2. Place the marinated pork slices in the steaming bowl in the form of book pages, then put the bowl into a steamer to steam over a high flame until cooked through. Remove from the steamer and turn the steaming bowl upside down to transfer the contents onto a serving dish.

NOTES

Suanzha Pork means sour preserved pork. It is a common way to cook Suanzha Pork in Anyue's countryside. The flavor is mainly influenced by one of the preparation steps—the time length sealed in a pottery and the temperature. In spring time, it is usually about 20 days. In summer or winter, the time could be slightly shortened or extended. This dish has bright color, lingering aroma, savory and sweet taste, fatty but not greasy pork. This cooking way can also be used for Suanzha Fish and Suanzha Beef, etc.

43 荷叶旱蒸滑肉
STEAMED HUAROU PORK WITH LOTUS LEAVES

食材配方

带皮五花肉200克
料酒5毫升
味精1克
红薯淀粉100克
临江寺豆瓣25克
姜米5克
刀口花椒5克
荷叶2张

制作工艺

1 将带皮五花肉切成片，加入临江寺豆瓣、料酒、姜米、味精、刀口花椒拌匀，分散放入不锈钢方盘，在每片肉上分别撒上红薯淀粉，用沸水浇淋定型。

2 荷叶洗净，放入小笼垫底，将定型的肉片放入，再盖上一张荷叶，用旺火蒸熟即成。

评鉴

安岳滑肉历史悠久，声誉远播，可煮、可蒸。荷叶旱蒸滑肉是在旱蒸滑肉的基础上使用荷叶包裹蒸制而成，既能保留滑肉透亮劲道的特点，又能突出荷叶的清香。

INGREDIENTS

200g pork belly with skin attached; 25g Linjiangsi chili bean paste; 5ml Shaoxing cooking wine; 5g ginger, finely chopped; 1g MSG; 5g Sichuan pepper, roasted and chopped; 100g sweet potato starch; 2 lotus leaves

PREPARATION

1. Slice the pork belly, mix well with Linjiangsi chili bean paste, Shaoxing cooking wine, ginger, MSG, Sichuan pepper, and lay separately on a stainless square tray. Sprinkle sweet potato starch over each slice of pork, and pour boiling water over them to fix the shape.

2. Rinse the lotus leaves, and place one leaf on the bottom of the steamer. Stack the pork slices on the lotus leaf, and cover with the other leaf. Steam over high heat till cooked through.

NOTES

Anyue Huarou, a renowned, time-honored local specialty, is prepared by boiling or steaming pork that has been coated with sweet potato starch. This dish is a modified version of Huarou, retaining the texture of Huarou and highlighting the fragrance of lotus leaves.

44 脆皮粉蒸肉
STEAMED PORK WITH CRISPY SKIN

食材配方

五花肉350克	糯米100克
香米120克	陈皮3克
八角3克	花椒3克
食盐2克	春卷皮12张
花椒粉3克	醪糟汁20毫升
料酒10毫升	姜米5克
葱米10克	蒜米5克
白糖4克	酱油4毫升
南乳汁10毫升	鸡精2克
临江寺豆瓣4克	鸡蛋液30毫升

制作工艺

1 锅置火上，放入糯米、香米炒至微黄，下陈皮、八角、花椒炒至金黄色，倒出晾凉，加工成粗米粉。

2 五花肉入锅中煮断生后捞出，切成长约8厘米、宽约5厘米、厚约0.4厘米的片，依次加入料酒、食盐、醪糟汁、白糖、姜米、葱米、蒜米、花椒粉、临江寺豆瓣、酱油、南乳汁、鸡精拌匀，再加入米粉拌匀，入笼蒸60分钟至肉软熟后出笼。

3 将粉蒸肉放在春卷皮上，卷裹成长方形，用鸡蛋液封口，入150℃的油中炸至金黄色捞出，装盘即成。

评鉴

脆皮粉蒸肉是在传统粉蒸肉的基础上创新而成。成菜色泽金黄，外酥内嫩，酥中带糯，糯中带甜，椒麻味浓郁。

INGREDIENTS

350g pork belly; 100g glutinous rice; 120g long grain rice; 3g dried orange peels; 3g star anise; 3g Sichuan pepper; 2g salt; 12 pieces of spring roll peels; 3g ground Sichuan pepper; 20ml fermented glutinous rice juice; 10ml Shaoxing cooking wine; 5g ginger, finely chopped; 10g spring onions, finely chopped; 5g garlic, finely chopped; 4g sugar; 4ml soy sauce; 10ml fermented doufu juice; 2g chicken essence; 4g Linjiangsi chili bean paste; 30ml beaten egg

PREPARATION

1. Place the wok on the stove. Dry roast glutinous rice and long grain rice until light yellow. Blend in dried orange peels, star anise and Sichuan pepper and fry till golden yellow. Remove the rice from the wok and leave to cool. Make coarse rice flour out of the rice.

2. Boil the pork belly till just cooked, and slice into pieces about 8cm long, 5cm wide and 0.4cm thick. Blend in Shaoxing cooking wine, salt, fermented glutinous rice juice, sugar, ginger, spring onions, garlic, ground Sichuan pepper, Linjiangsi chili bean paste, fermented doufu juice, chicken essence and rice flour. Steam for 60 minutes till the pork belly is soft and cooked through.

3. Roll the steamed belly pork with spring roll peels into the shape of cuboid and seal with egg juice. Deep-fry the rolls in 150℃ cooking oil till golden brown. Transfer to a serving dish.

NOTES

Steamed Pork with Crispy Skin is made based on the traditional Steamed Pork Belly. It is golden brown and tastes crispy, tender, sweet and spicy.

45 米粑粑滑肉汤
GLUTINOUS RICE CAKES IN HUAROU PORK SOUP

食材配方

里脊肉200克　老咸菜50克

米粑粑150克　食盐3克

料酒5毫升　姜米5克

鸡精2克　味精2克

葱花5克　鲜汤1000毫升

红薯粉300克　清水200毫升

制作工艺

1. 老咸菜切成颗粒；里脊肉切成薄片，加入食盐、料酒、鸡精、味精、红薯粉、清水拌匀。

2. 锅置火上，加入鲜汤、咸菜粒烧沸，将肉片分散放入后煮熟成滑肉，放入米粑粑、食盐、味精、鸡精烧沸，装入汤碗，撒上葱花即成。

评鉴

此菜的滑肉采用安岳特产红薯粉挂糊制成，色泽晶莹透亮，口感嫩滑劲道。在滑肉汤中加入米粑粑、老咸菜成菜，汤清味鲜，肉质嫩滑爽口，米粑粑软糯醇香，咸菜香浓，乡土气息突出

INGREDIENTS

200g tenderloin; 50g pickles; 150g glutinous rice cakes; 3g salt; 5ml Shaoxing cooking wine; 5g ginger, finely chopped; 2g chicken essence; 2g MSG; 5g spring onions, finely chopped; 1,000ml stock; 300g sweet potato starch; 200ml water

PREPARATION

1. Finely chop the pickles. Slice the tenderloin, add salt, Shaoxing cooking wine, chicken essence, MSG, sweet potato starch and water and mix well.

2. Add stock and pickles in the wok and boil. Place the meat in the wok and cook until done. Add the glutinous rice cakes, salt, MSG, chicken essence and boil. Serve at once with spring onions.

NOTES

Glutinous Rice Cakes in Huarou Pork Soup is made with the sweet potato starch of Anyue. It is crystal clear and tastes fresh and delicious. With glutinous rice and pickles in the broth, the soup is fresh, the meat is tender and smooth, the glutinous rice is soft and mellow, and the pickles are delicate, which shows obvious countryside flavors.

46 圣贤思乡肘
SHENGXIAN PORK KNUCKLE

食材配方

猪肘1个（约600克）　青尖椒100克
小米辣100克　　　　盐菜50克
姜米10克　　　　　　蒜米15克
葱花20克　　　　　　食盐3克
味精2克　　　　　　　芝麻油3毫升

制作工艺

1. 青尖椒、小米辣分别切成小圈；盐菜切成小颗粒。

2. 猪肘入锅煮熟，捞出沥干水分，入190℃的油锅中炸至表面金黄时捞出，再放入卤水锅中卤至软熟，捞出晾凉后去骨，用保鲜膜包裹定型，然后切成片，定碗，入笼蒸至软熟，出笼装盘。

3. 锅置火上，放油烧至120℃，放入姜米、蒜米、青尖椒、小米辣炒香，再入盐菜炒香，最后入食盐、味精、葱花、芝麻油炒匀，浇在肘子上即成。

评鉴

相传资阳三贤之一的苌弘出川后思念家乡味、家乡情，后人因此创制了此菜，收录在苌弘文化宴中。此菜色泽自然，质地软糯，味道咸鲜香辣。

INGREDIENTS

1 pork knuckle (about 600g); 100g green chili peppers; 100g bird's eye chilies; 50g pickles; 10g ginger, finely chopped; 15g garlic, finely chopped; 20g spring onions; 3g salt; 2g MSG; 3ml sesame oil

PREPARATION

1. Finely chop the green chili peppers, bird's eye chilies and pickles.

2. Boil the knuckle in the pot, remove from the wok and spiced drain. Deep fry the knuckle in the oil of 190°C till golden yellow, and boil in the broth till soft. Remove the knuckle from the wok and remove its bones. Wrap up with preservative film. Slice and place the meat in a bowl, and steam till soft and cooked through. Place in a plate.

3. Heat the oil in the wok to 120°C, and stir fry ginger, garlic, green chili peppers and bird's eye chilies. Add the pickles, salt, MSG, spring onions and sesame oil, and stir well. Sprinkle over the knuckle.

NOTES

It is recorded that as one of the Three Sages, Chang Hong missed the food of his hometown, Shengxian Pork Knuckle was then made and has been included in the Changhong cultural feast. This dish is natural and tastes soft, glutinous, salty and spicy.

47 大刀烧白
STEAMED SLICED PORK BELLY

食材配方

五花肉500克 豆芽150克
碎米芽菜100克 干辣椒节5克
香葱节5克 姜片5克
大葱段10克 食盐5克
酱油10毫升 味精2克
醪糟5克 白糖2克
胡椒粉1克 花椒1克
糖色30克 食用油30毫升
清水50毫升

制作工艺

1 五花肉切成长约10厘米、宽约8厘米的大块，入锅煮熟，捞出后晾干水分，趁热在肉皮抹上糖色，入200℃的油中炸至表皮起皱、呈棕红色时捞出，入冷水中浸泡，再切成长约8厘米、厚约1厘米的片。

2 将食盐、酱油、胡椒粉、味精、醪糟、白糖、糖色、清水入碗调成味汁。

3 将肉片摆放在蒸碗里呈一封书形，再放上碎米芽菜、花椒、姜片、大葱段，淋上味汁，上笼蒸2小时至肉软熟。

4 锅置火上，放入豆芽炒熟，入盘中垫底，再将蒸碗翻扣在豆芽上，放入炸香的干辣椒、香葱节即成。

评鉴

此菜在咸烧白的基础上演变而成，质地软糯，味道咸鲜浓厚，肥而不腻。

INGREDIENTS

500g pork belly; 150g bean sprouts; 100g yacai, finely chopped; 5g dried chilies, segmented; 5g chives, segmented; 5g ginger, sliced; 10g spring onions, segmented; 5g salt; 10ml soy sauce; 2g MSG; 5g fermented glutinous rice; 2g sugar; 1g pepper; 1g Sichuan pepper; 30g caramel; 30ml cooking oil; 50ml water

PREPARATION

1. Cut the pork belly into chunks about 10cm long and 8cm wide. Boil in the wok to be well done, remove and drain. Smear the pork skin with caramel and deep-fry in 200℃ till the skin becomes crispy and reddish brown. Remove the pork, soak in cold water and slice into pieces about 8cm long and 1cm thick.

2. Make sauce out of salt, soy sauce, pepper, MSG, fermented glutinous rice, sugar, caramel and water.

3. Place the sliced pork in the steaming bowl like an open book. Cover the pork with yacai, Sichuan pepper, ginger, and spring onions. Sprinkle the sauce and steam for 2 hours till the pork is soft.

4. Stir fry the bean sprouts and place on the dish. Cover the bean sprouts with the steamed pork. Serve at once with fried dried chili and chives.

NOTES

Steamed Sliced Pork Belly is developed based on Salty Steamed Pork Belly. It tastes soft, sticky, salty, fatty but not greasy.

48 松露狮子头
LION'S HEAD MEATBALLS WITH TRUFFLE

食材配方

五花肉600克
松露20克
水发香菇100克
枸杞10颗
食盐8克
胡椒粉1克
水淀粉50克
瓢儿白100克
冬笋100克
马蹄100克
鸡蛋清50克
料酒10毫升
味精2克

制作工艺

1. 五花肉切成约0.5厘米大的小颗粒；冬笋、水发香菇、马蹄分别切成小颗粒；松露洗净，切成薄片。

2. 五花肉粒入盆，放入冬笋、香菇、马蹄、鸡蛋清、食盐、料酒、胡椒粉、水淀粉搅匀成肉馅，制成直径4厘米的肉圆子，入汤锅中煨至软熟入味后捞出，分别放入紫砂盅内，加入松露片、枸杞、食盐、味精，入笼蒸20分钟取出，放入焯水后的瓢儿白即成。

评鉴

狮子头各地都有，此菜选用的是雁江区独特的伍隍猪肉制作而成，肉香味浓、口感好，配上松露片，别具一格。

INGREDIENTS

600g pork belly; 100g bok choy; 20g truffle; 100g winter bamboo shoots; 100g shiitakes; 100g Chinese water chestnuts; 10 wolfberries; 50g egg white; 8g salt; 10ml Shaoxing cooking wine; 1g pepper; 2g MSG; 50g batter

PREPARATION

1. Dice the pork belly into about 0.5cm small cubes; chop the winter bamboo shoots, shiitakes, Chinese water chestnuts respectively; Rinse the truffle, cut into thin slices.

2. Add the pork belly into a large bowl, add winter bamboo shoot, shiitakes, water chestnuts, egg white, salt, Shaoxing cooking wine, pepper and batter and mix together, stir well. Shape the mixture into large meatballs about 4cm in diameter. Add into a stock pot to simmer till cooked through and flavored, ladle out and transfer to purple clay cups separately, add truffle slices, wolfberries, salt, MSG, and then steam in a steamer for 20 minutes, add blanched bok choy then serve.

NOTES

This dish is localized by using a local specialty—the Wuhuang pork, which endows this dish with strong meat aroma and pleasant taste. With added truffle slices, this dish has a unique flavor.

食材配方

鸡蛋2个　　　　　碎肉150克
前夹肉200克　　　水发黄花100克
海带丝100克　　　胡萝卜丝50克
菜心150克　　　　姜葱水30毫升
食盐6克　　　　　味精2克
料酒10毫升　　　　红薯粉50克
鲜汤200毫升　　　水淀粉10克
清水30毫升

制作工艺

1. 鸡蛋打散搅匀，入锅煎成蛋皮；碎肉加入食盐、料酒、姜葱水、水淀粉调匀成馅；前夹肉切条，加入红薯粉、清水拌匀，入锅炸熟后装碗，再入笼蒸至软熟。

2. 蛋皮平铺，放入肉馅，中间分别放上海带丝、胡萝卜丝卷成如意卷，入笼蒸熟，取出晾凉，再切成长约1.5厘米的节。

3. 取一蒸碗，先放入蛋卷节，再放入酥肉、水发黄花、海带丝，加入鲜汤、食盐，入笼蒸15分钟后取出，滗出汤汁，翻扣在碗中，四周围上焯水后的菜心即成。

4. 锅置火上，加入原汤汁、食盐、味精烧沸，浇淋在蛋卷上即成。

评鉴

镶碗是四川百姓人家年夜饭的重头戏，少不了酥肉、蛋卷、黄花等食材，但各地选料略有差异。此菜质地酥软细腻，味道咸鲜清香。

INGREDIENTS

2 eggs; 150g pork mince; 200g pork shoulder; 100g long yellow daylilies, water-soaked; 100g seaweeds, shredded; 50g carrots, shredded; 150g choy sums; 30ml ginger-and-scallion-flavored juice; 6g salt; 2g MSG; 10ml Shaoxing cooking wine; 50g sweet potato starch; 200ml stock; 10g batter; 30ml water

PREPARATION

1. Whisk the egg well, fry in the pan into omelets. Blend the pork mince well with salt, Shaoxing cooking wine, ginger-and-scallion-flavored juice and batter to make the filling. Slice the pork shoulder, blend with sweet potato starch and water, then deep-fry until cooked through. Remove to a bowl and steam until soft.

2. Place the filling on omelets, add seaweeds and carrots on each piece at the same time. Roll up the omelets, and steam until cooked through. Remove and cool down, then cut it into about 1.5cm long segments.

3. Put the omelet roll segments in a large bowl first, then add the pork, long yellow daylilies, seaweeds, and blend in stock and salt at last. Steam for 15 minutes and remove. Skim the stock, transfer the content upside-down in another large bowl, add blanched choy sums around.

4. Heat oil in a wok, add the stock, salt and MSG to boil, then pour over the egg omelets and done.

NOTES

Xiangwan Bowl is a common Spring Festival dish for every household in Sichuan. Deep-fried pork, egg omelets, long yellow daylilies are the regular ingredients for this dish. However, it depends on regions. This dish has fresh fragrance smell, savory and tender taste.

49 镶碗

50 柚香排骨
POMELO-FLAVORED FRIED SPARERIBS

食材配方

猪排500克　　鲜柚皮300克
食盐5克　　　葱段20克
姜片10克　　　料酒10毫升
鸡蛋清60克　　味精1克
淀粉20克　　　白芝麻10克
食用油1 000毫升（约耗100毫升）

制作工艺

1. 将猪排斩成长约4厘米的段，加入食盐、姜片、葱段、料酒码味，入180℃的油中炸至成熟、酥香时捞出。
2. 鲜柚皮去掉青皮，切成长约4厘米的片，入锅焯水后捞出，沥干水分，加入鸡蛋清、淀粉拌匀，入160℃的油中炸至成熟、酥香时捞出。
3. 锅置火上，放油烧至120℃，入排骨、柚皮、食盐、白芝麻、味精炒匀，起锅装盘即成。

评鉴

此菜选用安岳特产的通贤柚皮为食材制成。通贤柚是国家地理标志产品，有生津止渴、增进食欲、提神醒脑的作用。此菜将柚皮裹蛋清淀粉糊后油炸，再与排骨同炒，具有外酥内嫩、柚香浓郁的特点。

INGREDIENTS

500g spareribs; 300g fresh pomelo rinds; 5g salt; 20g spring onions, segmented; 10g ginger, sliced; 10ml Shaoxing cooking wine; 60g egg white; 1g MSG; 20g cornstarch; 10g white sesames; 1,000ml cooking oil (about 100ml to be consumed)

PREPARATION

1. Chop the spareribs into about 4cm-long segments and season with salt, ginger, spring onions and Shaoxing cooking wine. Then heat the oil to 180℃ and deep fry the spareribs till cooked through and crispy.

2. Remove the green peels of fresh pomelo rinds and cut into slices of 4cm long. Blanch the rinds and drain. Then add egg white and cornstarch and mix them well. Heat the oil to 160℃ and deep-fry till cooked through and crispy.

3. Heat the oil to 120℃. Add spareribs, pomelo rinds, salt, white sesames and MSG, stir fry and transfer to the serving dish.

NOTES

The major ingredient of this dish is the rinds of Tongxian pomelo which is a specialty of Anyue. It is also a product with national geographical indication which can promote the secretion of saliva and quench thirst, whet people's appetite, refresh people and invigorate the function of brain. The rind slices are deep-fried after being wrapped in the batter made of egg white and cornstarch and then stir-fried with spareribs. Therefore it tastes crispy outside while soft inside with a strong scent of pomelo.

51 滋补莲藕汤
LOTUS ROOT CASSEROLE

食材配方

鲜藕500克　　排骨300克
棒骨500克　　党参20克
大枣20克　　　枸杞5克
食盐10克　　　姜片15克
葱段20克　　　料酒20毫升
胡椒粉1克　　 味精2克

制作工艺

1 排骨斩成长约6厘米的段，焯水；棒骨敲破，焯水；鲜藕去皮，切成大的滚料块；党参用清水浸泡软，切成长约4厘米的段。

2 砂锅置火上，入排骨、棒骨，掺入清水烧沸，去掉浮沫，放入姜片、葱段、料酒、胡椒粉，改用小火炖1小时，入鲜藕、党参、大枣、枸杞继续炖1小时，捞出棒骨，放入食盐、味精，装盘即成。

评鉴

莲藕富含维生素C及矿物质，有促进新陈代谢、防止皮肤粗糙的作用。此菜将莲藕、排骨与具有气血双补作用的党参、大枣、枸杞同炖，味道咸鲜，滋补作用更佳。

INGREDIENTS

500g fresh lotus roots; 300g pork ribs; 500g pork bones; 20g codonopsis; 20g Chinese dates; 5g wolfberries; 10g salt; 15g ginger, sliced; 20g spring onions, segmented; 20ml Shaoxing cooking wine; 1g pepper; 2g MSG

PREPARATION

1. Cut the pork ribs into about 6cm segments and blanch. Crack the pork bones and blanch. Peel the lotus roots and cut into big rolling chunks. Soak the codonopsis in the water until soft and then cut into about 4cm segments.

2. Heat the pot and add the pork ribs, pork bones and water until they are boiling. Skim off the scum and add ginger slices, spring onions segments, Shaoxing cooking wine and pepper, stew them over a low flame for 1 hour and then add lotus roots, codonopsis, Chinese dates and wolfberries, continue to stew for another 1 hour. Then remove the pork bones and add salt and MSG.

NOTES

The lotus roots are rich in Vitamin C and mineral substances, which helps to accelerate the process of metabolism and avoid rough skin. The lotus roots are stewed with the pork ribs, codonopsis, Chinese dates and wolfberries, which are all beneficial for blood circulation. It is nutritious and tastes salty and savory.

52 竹荪肝膏汤
LIVER PASTE SOUP WITH BAMBOO FUNGUS

食材配方

猪肝200克　　　竹荪10克
鸡蛋清100克　　姜片20克
葱节30克　　　胡椒粉1克
食盐5克　　　　味精2克
鸡汤100毫升　　清汤1 250毫升

制作工艺

1. 竹荪用温水泡涨，切成长约2厘米的节，焯水后用清汤浸泡备用；姜片、葱节、胡椒粉用清水浸泡制成姜葱胡椒水。

2. 猪肝去筋，捶成细蓉，入碗内加鸡汤调匀，用箩筛滤去肝渣；碗内肝汁加入鸡蛋清、姜葱胡椒水、食盐、味精调匀，入笼用中气蒸12分钟凝结成肝膏。

3. 锅置火上，加入清汤、竹荪烧沸，倒入汤碗，放入肝膏即成。

评鉴

此菜是川菜传统经典菜肴，肝膏细嫩，竹荪脆滑，汤色清澈，味道咸鲜醇厚，清淡爽口。

INGREDIENTS

200g pork liver; 10g veiled bamboo fungus; 100g egg white; 20g ginger, sliced; 30g spring onions, segmented; 1g pepper; 5g salt; 2g MSG; 100ml chicken broth; 1,250ml consommé

PREPARATION

1. Put the bamboo fungi into the warm water till expand, then cut them into segments of 2cm long. Blanch them and soak in the consommé. Soak the ginger slices, spring onions and pepper in water to make the ginger-scallion-pepper flavored juice.

2. Dislodge the fascia of the pork liver and then finely chop the liver. Transfer to a bowl, add chicken broth and mix well. Filter out the liver dregs with a sieve. Add egg white, ginger-scallion-pepper flavored juice, salt and MSG in the bowl of liver juice. Put the bowl in the steamer to steam for 12 minutes till the live juice congeals into liver paste.

3. Heat the pot, add consommé and bamboo fungus till boiling, transfer to the serving dish and add the liver paste.

NOTES

The dish is one of the traditional and classic Sichuan cuisines. The liver paste is soft and tender while the veiled bamboo fungi are crispy and smooth. The soup is transparent with salty, savory and refreshing flavor.

53 椒盐猪肝
PEPPER SALT PORK LIVER

食材配方

猪肝300克
食盐5克
水淀粉30克
花椒粉6克
食用油500毫升（约耗60毫升）
葱花15克
料酒10毫升
味精2克
白糖20克

制作工艺

1 猪肝切片，加入食盐、料酒、水淀粉拌匀。

2 锅置火上，放油烧至150℃，放入猪肝炸至定型后捞出，待油温回升至200℃时，入锅复炸至外酥内软，滗去多余的油，再放入葱花、食盐、花椒粉、白糖、味精翻炒均匀，起锅装盘即成。

评鉴

此菜的技术要求较高，需要掌握好两次油炸的温度和火候，成菜具有味道咸鲜麻香、口感软糯的特点。

INGREDIENTS

300g pork liver; 15g green onions, finely chopped; 5g salt; 10ml Shaoxing cooking wine; 30g batter; 2g MSG; 6g ground Sichuan pepper; 20g sugar; 500ml cooking oil (about 60ml to be consumed).

PREPARATION

1. Slice the pork livers, add salt, Shaoxing cooking wine and batter and mix well.

2. Heat the cooking oil in the wok to 150 °C. Fry the sliced pork liver and remove from the wok after their shape has formed. When the oil temperature rises to 200 °C, fry the liver to be crispy. Remove the excessive oil, add green onions, salt, ground Sichuan pepper, sugar, and MSG and stir fry before it's done.

NOTES

The technical requirements of this dish are high, and it is necessary to delicately handle the temperature and heat for frying of two times. The fried pork liver tastes salty, fresh and soft.

食材配方

安岳柠檬蜜饯20克　　鸡蛋黄8个

水淀粉20克　　白糖30克

食盐3克　　沸水150毫升

化猪油100毫升

制作工艺

1 柠檬蜜饯剁细；鸡蛋黄入碗，加入水淀粉、白糖、食盐调匀，加入沸水，调制成蛋浆，再加入柠檬蜜饯调匀。

2 锅置火上，放化猪油烧至120℃，倒入蛋浆炒至凝固且成熟，起锅装盘即成。

评鉴

此菜色泽金黄，味道咸甜，柠檬香浓，质地滑嫩，入口即化。

INGREDIENTS

20g preserved Anyue lemon; 8 egg yolks; 20g batter; 30g sugar; 3g salt; 150ml boiling water; 100ml lard

PREPARATION

1. Chop the preserved lemon finely. Mix egg yolks well with batter, sugar and salt, then add boiled water to make egg batter. Add preserved lemon and mix well.

2. Heat the lard in a wok until 120℃, add egg batter and stir until curdled and well-cooked, remove to a serving dish.

NOTES

This dish has golden color, rich lemon fragrance, delicate texture, salty and sweet taste, easily melting in mouths.

54 柠香熘黄菜 LEMON-FLAVOR STIR FRIED EGG YOLKS

食材配方

带皮野猪五花肉400克	水发干豇豆150克
蒜片50克	泡辣椒末30克
泡姜米20克	泡辣椒节10克
蒜节10克	料酒100毫升
红油50毫升	食盐5克
水淀粉20克	味精2克
花椒5克	糖色5克
辛香料3克	鲜汤1 000毫升
食用油200毫升	

制作工艺

1 野猪肉烧皮后洗净，切成约2厘米见方的丁，入锅焯水备用。

2 干豇豆切成长约2.5厘米的节，入锅焯水备用。

3 锅置火上，放油烧至120℃，放入泡辣椒末、泡姜米、花椒炒香且油呈红色，掺入鲜汤烧沸，沥去料渣，放入野猪肉、豇豆节、食盐、糖色、辛香料、料酒烧1小时，放入蒜片，改小火再烧20分钟至汤少时加入泡辣椒节、蒜节、红油、味精烧沸，入水淀粉收汁浓稠，起锅装盘即成。

评鉴

这里的野猪是指农家放养的猪，其肉的胶质重、脂肪少、口感好，是理想的绿色健康食材。野猪肉与豇豆一起烧制，相互融合，香味浓郁，口感软糯不腻。

INGREDIENTS

400g free-range pork with skin attached; 150g water-soaked dry long beans; 50g garlic, sliced; 30g pickled chilies, finely chopped; 20g pickled ginger, finely chopped; 10g pickled chilies, segmented; 10g green garlic, segmented; 100ml Shaoxing cooking wine; 50ml chili oil; 5g salt; 20g batter; 2g MSG; 5g Sichuan pepper; 5g caramel; 3g spices; 1,000ml stock; 200ml cooking oil

PREPARATION

1. Scorch the pork skin and rinse. Cut the pork into about 2cm cubes, and blanch.
2. Cut the long beans into about 2.5cm long sections, and blanch.
3. Heat the oil in a wok to 120°C. Add pickled chilies, ginger and Sichuan pepper, stir fry till aromatic and the oil becomes reddish. Pour in the stock, bring to a boil and scum. Add the pork, long beans, salt, caramel, spices and Shaoxing cooking wine, and braise for 1 hour. Add the garlic, and continue to simmer over low heat for another 20 minutes before blending in the pickled chilies, green garlic, chili oil and MSG. Bring to a boil, pour in the batter to thicken the sauce. Transfer to a serving dish.

NOTES

The free-range pork used here is a green, healthy food, featuring low fat and high collagen. The pork and long beans absorb flavors from each other, presenting a delicacy with tender texture.

55 干豇豆烧野猪肉
BRAISED FREE RANGE PORK WITH DRIED LONG BEANS

食材配方

带皮黑山羊肉500克	白萝卜600克
香菜10克	仔姜50克
青尖椒20克	小米辣10克
临江寺豆瓣30克	干辣椒8克
花椒5克	八角3克
桂皮5克	姜米10克
蒜米10克	啤酒20毫升
清水1 000毫升	鲜汤1 000毫升
食盐15克	味精2克
辣椒油100毫升	

制作工艺

1. 带皮黑山羊肉斩成约4厘米见方的块；白萝卜切成大三角形块；青尖椒、小米辣切成段。

2. 锅置火上，放入辣椒油烧至150℃，放入羊肉块煸炒至香，入八角、桂皮、临江寺豆瓣、姜米、蒜米、干辣椒、花椒炒香，再加入啤酒和清水烧沸，改用小火烧至熟软后，入青尖椒、小米辣、仔姜、味精烧沸，起锅装入盘的中间。

3. 锅中掺入鲜汤，放入白萝卜、食盐煮至熟软，放在羊肉四周，撒上香菜即成。

评鉴

乐至黑山羊是资阳市乐至县特产，全国农产品地理标志产品，具有味香美、不膻、肉质细嫩等优点，配搭萝卜、青椒成菜，色泽自然，味道咸鲜、略带麻辣，香味浓郁。

INGREDIENTS

500g black goat meat with skin; 600g white radishes; 10g coriander; 50g tender ginger; 20g green chili peppers; 10g bird's eye chilies; 30g Linjiangsi chili bean paste; 8g dried chilies; 5g Sichuan pepper; 3g star anise; 5g cinnamon; 10g garlic, finely chopped; 10g ginger, finely chopped; 20ml beer; 1,000ml water; 1,000ml stock; 15g salt; 2g MSG; 100ml chili oil

PREPARATION

1. Cut the black goat meat with skin into about 4cm chunks. Cut the white radishes into triangles. Cut green chili pepper and bird's eye chilies into segments.

2. Heat the chili oil in the wok to 150℃. Fry the meat and blend in star anise, cinnamon, Linjiangsi chili bean paste, ginger, garlic, dried chilies and Sichuan pepper and fry. Add the beer and water in the pot and boil. After the soup is boiling, turn the stove to low heat till the meat is well done. Add green chili peppers, bird's eye chilies and MSG and boil. Transfer the meat to the middle of a plate.

3. Add the stock, white radishes and salt in the pot and boil till the radishes are soft. Sprinkle coriander around the meat on the plate.

NOTES

Braised black goat meat is a special dish of Lezhi County of Ziyang City. The goat is famous in China and its meat tastes delicious and tender. With white radishes and green chili pepper, the dish is fresher and spicier.

56 黄焖烫皮黑山羊
BRAISED BLACK GOAT MEAT

57 仔姜牛肉丝
BEEF SLIVERS WITH TENDER GINGER

食材配方

牛里脊肉200克　　仔姜100克
红甜椒40克　　　甜面酱10克
水淀粉20克　　　味精2克
食盐5克　　　　　料酒10毫升
鲜汤25毫升　　　食用油100毫升

制作工艺

1 牛里脊肉切成二粗丝，加入食盐、料酒、水淀粉拌匀；仔姜、红甜椒分别切成长约5厘米的丝。

2 将食盐、料酒、味精、水淀粉、鲜汤入碗调成芡汁。

3 锅置火上，放油烧至180℃，放入牛肉丝炒散至熟，入甜面酱炒散，入仔姜、红甜椒炒熟，倒入芡汁，收汁亮油后起锅装盘即成。

评鉴

此菜选用安岳县鼎新乡出产的仔姜，与牛肉丝炒制成菜，色泽美观，肉质滑嫩，仔姜脆爽，味道咸鲜微辣。

INGREDIENTS

200g beef sirloin ; 100g tender ginger; 40g red bell peppers; 10g fermented flour paste; 20g batter; 2g MSG; 5g salt ;10ml Shaoxing cooking wine; 25ml stock; 100ml cooking oil

PREPARATION

1. Cut the beef sirloin into medium slivers, then add salt, Shaoxing cooking wine and batter and mix them well. Cut the tender ginger and red bell peppers into about 5cm slivers.

2. Add salt, Shaoxing cooking wine, MSG, batter and stock into a bowl and mix them as the thickening sauce.

3. Heat the oil to 180℃ in a wok, and then add beef slivers and stir fry till cooked through. Then add fermented flour paste, tender ginger and red bell peppers, stir fry till cooked through. Add the thickening sauce and stir fry till the sauce is reduced. Then transfer to the serving dish.

NOTES

This dish is made from tender ginger, freshly produced in Dingxin Township, Anyue County, and beef slivers. The color is beautiful, the beef is smooth and tender, and the ginger is crispy. It tastes salty, savory and a little spicy.

58 松茸牦牛排

食材配方

牦牛排500克　鲜松茸50克
小金瓜50克　干葱50克
大蒜50克　姜片20克
孜然粒3克　食盐5克
蚝油20毫升　东古一品鲜10毫升
鸡粉5克　鲜汤500毫升
食用油1000毫升（约耗60毫升）

制作工艺

1 将牦牛排切成约3厘米见方的方块，入160℃的油锅炸至表皮金黄时捞出；松茸用黄油煎至两面香；小金瓜切块，焗烤至熟。

2 锅置火上，放油烧至120℃，放入干葱、大蒜、姜片炸香，入孜然粒炒香，再入牦牛排、蚝油、食盐、东古一品鲜、鸡粉、鲜汤，用小火慢煲至牛排软熟，最后入金瓜、鲜松茸，收汁浓稠后装盘即成。

评鉴

牦牛肉是高原特产优质食材，高蛋白、低脂肪、低热量，富含多种氨基酸，营养价值远高于普通牛肉。此菜将牦牛肉和松茸用文火慢慢烹制，能最大程度保持肉的香味和弹性，味道咸鲜香浓。

SIMMERED YAK STEAK WITH PINE MUSHROOM

INGREDIENTS

500g yak steak; 50g fresh pine mushroom; 50g red peel pumpkin; 50g shallot; 50g garlic; 20g ginger, sliced; 3g cumin, finely chopped; 20ml oyster sauce; 10ml Donggu soybean sauce; 5g chicken essence; 500ml stock; 1,000ml cooking oil (about 60ml to be consumed)

PREPARATION

1. Cut the yak steak into cubes of about 3cm long and heat the oil to 160℃, put the yak steak in and deep fry till golden. Shallow fry the pine mushroom with butter till aromatic. Cut the red peel pumpkin into chunks and bake till they are cooked through.

2. Heat the oil to 120℃, add shallot, garlic and ginger slices to fry till aromatic, add cumin to fry till aromatic, then add yak steak, oyster sauce, Donggu soybean sauce, chicken essence and stock, simmer over a low flame till the steak is soft and cooked through. Add red peel pumpkin and fresh pine mushroom and continue to simmer till the sauce is reduced. Then transfer to the serving dish.

NOTES

Yak meat is the special high-quality food material in highland areas. It is rich in protein and various amino acid while low in fat and calorie, which has much higher nutritive value than the general beef. This dish is simmered by using yak and pine mushroom over a low flame, which can maintain the fragrance and elasticity of the meat. It is salty and savory in flavor.

59 洗手鲊牛肉
XISHOUZHA BEEF

食材配方

净牛肉300克　熟米粉150克
姜米20克　　泡姜米20克
临江寺豆瓣20克　食盐3克
料酒10毫升　香料6克
葱花10克　　白糖5克
猪油30毫升　菜籽油100毫升
清水500毫升

制作工艺

1. 牛肉切成约2厘米见方的大丁。

2. 锅置火上，放入猪油和菜籽油烧至180℃，先放入牛肉丁炒熟，再入临江寺豆瓣、姜米、泡姜米、香料、白糖、料酒、食盐炒香，掺入清水，烧至牛肉㸆软后，入熟米粉、葱花炒匀，起锅装盘即成。

评鉴

此菜具有浓厚的乡村气息，色泽红亮，质地软糯，味道咸鲜香辣。据说在安岳的一些地方，农村妇女下地干活，直到临近吃饭时才回家洗手、做饭，为尽快出菜，就将熟米粉放入烧肉的锅中与肉炒匀，即现鲊成菜。这种方法与用米粉裹肉后长时间发酵的鲊肉方法有相似之处，但也有不同，因此当地人将这种做鲊肉的方法称为"洗手鲊"。用此方法还可以制作洗手鲊肉、洗手鲊肥肠等菜。

INGREDIENTS

300g clean beef; 150g cooked ground rice; 20g ginger, finely chopped; 20g pickled ginger, finely chopped; 20g Linjiangsi chili bean paste; 3g salt; 10ml Shaoxing cooking wine; 6g spices; 10g spring onions, finely chopped; 5g sugar; 30ml lard; 100ml rapeseed oil; 500ml water

PREPARATION

1. Cut the beef into cubes of about 2cm long.

2. Heat the lard and rapeseed oil in the wok to 180℃. Add beef cubes and stir-fry. Then add Linjiangsi chili bean paste, ginger, pickled ginger, spices, sugar, Shaoxing cooking wine and salt, scramble till aromatic. Add water and braise the beef until it becomes soft. Add the cooked ground rice and spring onions and transfer to the serving dish.

NOTES

This dish has a strong countryside style with its bright red color, glutinous texture, and salty, fresh and spicy tastes. It is said that women in some rural areas in Anyue spent a lot of time working in the field and they could only wash their hands and cook until it was time for meal. In order to prepare dishes as soon as possible, they put cooked ground rice into the wok with meat and fried them well to be this dish. This method has something in common with the way of fermenting meat for a long time with ground rice wrapped while it also has its own characteristics. Therefore, the local people call this method as Xishouzha, which means make the dish after washing hands. This cooking method is also used to make Xishouzha pork and Xishouzha pork intestines.

60 酥炸桑叶配牛肉
DEEP-FRIED MULBERRY LEAVES WITH BEEF

食材配方

嫩桑叶100克　　安格斯牛肉300克
红酒30毫升　　　黑胡椒2克
黑胡椒汁100毫升　天妇罗粉100克
食盐5克　　　　　沙拉酱50克
清水70毫升
食用油2 000毫升（约耗100毫升）

制作工艺

1. 天妇罗粉加入清水、食盐调成糊；牛肉切成丁，加入黑胡椒、红酒码味。
2. 锅置火上，放油烧至150℃，放入挂了糊的桑叶，炸至酥脆时捞出，装入盘子的一端。
3. 平煎锅置火上，放油烧至180℃，放入牛肉丁煎熟，入黑胡椒汁炒匀时捞出，盛入盘子的另一端，配上沙拉酱味碟即成。

评鉴

此菜中西结合，本地时令桑叶撞上美味可口的西式牛肉，衍生出了美味佳肴，桑叶香脆，牛肉细嫩，层次丰富。

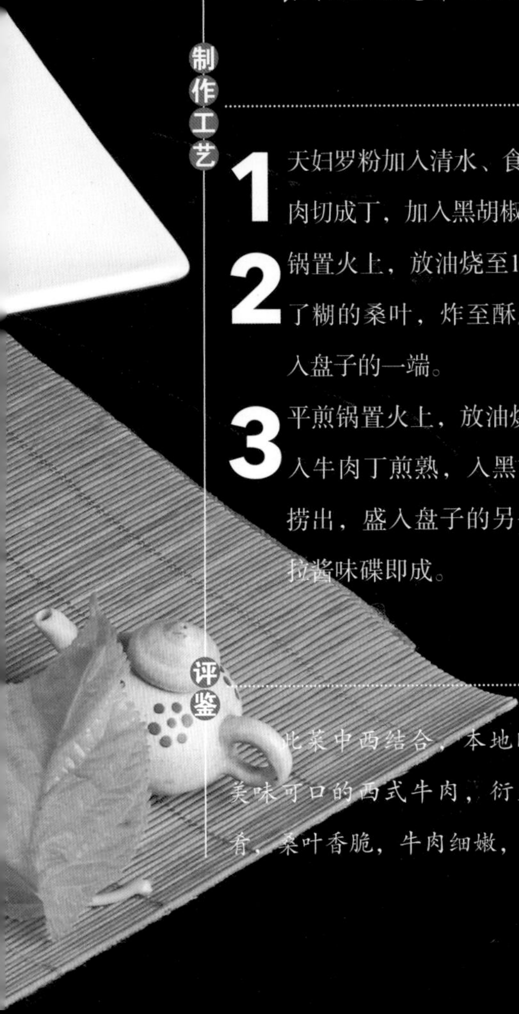

INGREDIENTS

100g mulberry leaves; 300g angus beef; 30ml red wine; 2g black pepper; 100ml black pepper sauce; 100g Tempura flour; 5g salt; 50g salad sauce; 70ml water; 2,000ml cooking oil (about 100ml to be consumed)

PREPARATION

1. Add Tempura flour, water, and salt to make paste. Cut the beef into chunks and add black pepper and red wine.

2. Heat the oil to 150℃ in a wok and add the mulberry leaves covered by the prepared Tempura paste. Deep-fry the mulberry leaves. Remove and put the leaves on one side of the plate when crispy.

3. Heat the oil to 180℃ in a pan and fry the beef chunks till them are cooked through. Add black pepper sauce and remove the beef chunks. Put them on the other side of the plate with the salad sauce dish.

NOTES

This dish combines both Chinese and the Western features. Local seasonal mulberry leaves together with the tasty western beef generate this delicious cuisine. The mulberry leaves are crispy and the beef is fresh and tender, making this dish well-structured.

INGREDIENTS

500g beef ribs; 6 pieces of edible rice paper; 200g bread crumbs; 2 eggs; 30ml butter; 20g cornstarch; 3,000ml pure water; 10g rock sugar; 50g coriander; 50g onions; 50g celery; 25g beef broth; 20g ginger; 30g dried chilies; 10g Sichuan pepper; 5g salt; 20ml Shaoxing cooking wine; 5g star anise; 5g bay leaves; 3g cinnamon; 3g caoguo herb; 1,000ml cooking oil (about 60ml to be consumed)

PREPARATION

1. Heat the butter in a pan to 160℃, fry the beef ribs until its surface golden brown; blend the egg and cornstarch well to make the egg batter.

2. Heat the oil in a wok to 150℃, stir-fry ginger, dried chilies, Sichuan pepper, Shaoxing cooking wine, star anise, bay leaves, cinnamon, caoguo herb, coriander, onions and celery until aromatic, add the pure water, beef broth, rock sugar, salt and beef ribs and simmer until the beef loosens away from bones, ladle out and remove the beef from bones.

3. Wrap the beef with rice paper into long rolls, coat with the egg batter first, then the bread crumbs, deep-fry in 130℃ oil until

61 雁江春暖
YANJIANG WARM SPRING / YANJIANG RIVER'S WARM SPRING

食材配方

牛肋骨500克	糯米纸6张
日本面包糠200克	鸡蛋2个
黄油30毫升	生粉20克
纯净水3 000毫升	冰糖10克
香菜50克	洋葱50克
芹菜50克	牛肉汁25克
生姜20克	干辣椒节30克
花椒10克	食盐5克
料酒20毫升	八角5克
香叶5克	桂皮3克
草果3克	食用油1 000毫升（约耗60毫升）

制作工艺

1. 平底锅置火上，放入黄油烧至160℃，入牛肋骨煎至表面金黄；鸡蛋和生粉调成全蛋淀粉糊。

2. 锅置火上，放油烧至150℃，放入生姜、干辣椒节、花椒、料酒、八角、香叶、桂皮、草果、香菜、洋葱、芹菜炒香，加入纯净水、牛肉汁、冰糖、食盐、牛肋骨，用小火煲至牛肉离骨，捞起晾干，剔下净肉。

3. 将牛肋骨肉用糯米纸包裹成长卷，粘全蛋淀粉糊，再粘上面包糠，入130℃的油中炸至色金黄、皮酥脆后捞出，切片装盘即成。

评鉴

此菜创意体现春暖花开、老农耕田的意境，牛肉片层次分明、立体感强，味道咸鲜，微带麻辣，质地酥脆化渣。

golden brown and crispy, slice and set on a serving plate.

NOTES

This dish is created to represent people's appreciation of spring blossom and farming of peasants. The beef slices have distinct layers, giving us a vivid look of spring fields. This dish is easily melting in mouth and has salty, savory, slightly spicy, crispy taste.

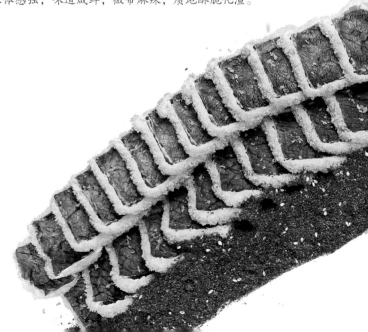

62 乡村羊腿
RURAL LAMB LEGS

食材配方

黑山羊前腿1个（约1 500克）　五花肉丁30克
青椒100克　　　　　　　　　红甜椒100克
香菇丁15克　　　　　　　　　食盐15克
辛香料10克　　　　　　　　　料酒15毫升
姜30克　　　　　　　　　　　蒜10克
大葱150克　　　　　　　　　 干辣椒5克
临江寺豆瓣20克　　　　　　　复合酱5克
鲜汤30毫升　　　　　　　　　鸡汁5毫升
烧椒碟子1个　　　　　　　　 秘制酱碟1个
食用油20毫升

制作工艺

1 食盐、辛香料炒香；青椒、红甜椒切成颗粒；取大葱100克切成长约4厘米的节，再用刀改成菊花形，入水浸泡成花葱。

2 黑山羊腿入食盐、辛香料、姜、大葱、料酒码味，腌制24小时，入卤水锅中卤3小时至羊腿软熟入味后捞出，沥干水分，入油锅炸至香酥后捞出装盘。

3 锅置火上，放油烧至150℃，入五花肉丁炒香，入干辣椒、临江寺豆瓣、姜、蒜、香菇丁炒香，再入鲜汤、鸡汁、青红椒丁、复合酱收汁，起锅淋在羊腿上，配上花葱、烧椒碟子、秘制酱碟成菜。

评鉴

此菜选用乐至黑山羊的前腿制作而成，豪放大气，外酥内嫩，味道咸鲜微辣，酱香突出，乡土气息浓郁。

INGREDIENTS

1 black goat foreleg (about 1,500g); 30g pork belly cubes; 100g green chilies; 100g red bell peppers; 15g shiitake mushroom cubes; 15g salt; 10g spices; 15ml Shaoxing cooking wine; 30g ginger; 10g garlic; 150g spring onions; 5g dried chilies; 20g Linjiangsi chili bean paste; 5g compound-flavored sauce; 30ml stock; 5ml chicken broth; 1 grilled chili sauce saucer; 1 homemade sauce saucer; 20ml cooking oil.

PREPARATION

1. Fry the salt and spices till aromatic. Chop the green chilies and red peppers. Cut the leek into 4cm segments. Then cut the leek segments into chrysanthemum shape and soak it in water.

2. Season the black goat foreleg with salt, spices, ginger, spring onions, and Shaoxing cooking wine for 24 hours. Boil the black goat foreleg in spiced broth for 3 hours till soft and tasty, then remove and drain the water. Deep-fry till crispy and aromatic.

3. Heat the oil in the wok to 150℃. Stir-fry the pork belly cubes till aromatic. Add dried chilies, Linjiangsi chili bean paste, ginger, garlic, and shiitake mushroom and stir fry till aromatic. Add stock, chicken broth, cubes of green chilies and red peppers, compound-flavored sauce. Stir fry till the sauce is reduced and pour over the prepared black goat foreleg, and serve with spring onions, the grilled chili sauce saucer and homemade sauce saucer.

NOTES

This dish is made of the foreleg of black goat raised in Lezhi County. It is crispy on the outside and soft on the inside. It tastes salty and a little spicy, with strong aroma of soybean sauce.

食材配方

黑山羊排400克	松茸片80克
黑胡椒5克	粗海盐5克
黄油20毫升	大蒜10克
莳萝5克	白洋葱200克
红甜椒50克	胡萝卜30克

制作工艺

1 黑山羊排加粗海盐、黑胡椒、莳萝、白洋葱、红甜椒、胡萝卜拌匀,抽真空腌制2小时;松茸切片。

2 锅置火上,加黄油烧热,放入黑山羊排、松茸片,再放入粗海盐、黑胡椒、莳萝、大蒜,将羊排煎至金黄焦香后出锅装盘即成。

评鉴

此菜以本地产的黑山羊排为食材,结合西式调料和制作方法,中菜西做,外香内嫩,黑胡椒味浓。

INGREDIENTS

400g lamb ribs; 80g pine mushroom; 5g black pepper; 5g sea salt; 20ml butter; 10g garlic; 5g dill; 200g white onions; 50g red bell peppers; 30g carrots

PREPARATION

1. Mix lamb ribs well with sea salt, black pepper, dill, white onions, red bell peppers, carrot and vacuum to marinate for 2 hours.

2. Heat the butter in a pan, add lamb ribs and pine mushroom slices, then add sea salt, black pepper, dill, onions to fry until golden brown then remove and transfer to a serving dish.

NOTES

The lamb ribs are a local product—black lambs' ribs. Applying western seasonings and cooking ways to this Chinese dish makes it have black pepper aroma, crispy skin and tender meat.

63 松茸香煎黑山羊
FRIED PINE MUSHROOMS WITH LAMB

64 BRAISED LAMB CHOPS WITH POTATOES

酱土豆焖羊排

食材配方

黑山羊排750克　　洋葱100克
胡萝卜200克　　　大蒜20克
姜片20克　　　　　黑胡椒汁30毫升
复合酱50克　　　　熟蛋黄1个
红豆腐乳20克　　　香菜杆20克
土豆500克　　　　 临江寺豆瓣20克
鲜汤500毫升　　　 食用油50毫升

制作工艺

1 羊排斩成段，焯水后过油；胡萝卜切成厚片。

2 锅置火上，放油烧至120℃，放入临江寺豆瓣、洋葱、大蒜、姜片、胡萝卜片、红豆腐乳炒香，再加入鲜汤、熟蛋黄、香菜杆、羊排、复合酱，用小火炖2小时，加入黑胡椒汁至收汁浓稠。

3 土豆挖成10个圆球形，加入食盐，入180℃的油锅中炸至金黄色捞出，入烧羊排的汁水略烧至成熟。

4 羊排装入盘中，周边放上土豆球即成。

评鉴

此菜精选乐至特产黑山羊的羊排，融入西餐制作方法，中西合璧，成菜色泽棕红，肉质滑嫩，味道咸鲜，微带麻辣，酱香浓郁。

INGREDIENTS

750g lamb chops; 100g onions; 200g carrots; 20g garlic; 20g ginger, sliced; 30ml black pepper sauce; 50g seasoned sauce; 1 cooked egg yolk; 20g fermented doufu; 20g coriander stems; 500g potatoes; 20g Linjiangsi chili bean paste; 500ml stock; 50ml cooking oil

PREPARATION

1. Cut the lamb chops into segments, boil and fry to be medium well. Slice the carrots into thick pieces.

2. Put the wok on the stove and heat the cooking oil in it to 120℃. Blend in Linjiangsi chili bean paste, onions, garlic, ginger, carrot and fermented doufu and fry. Add the stock, cooked egg yolk, coriander stems, lamb chops and seasoned sauce, and braise with low heat for 2 hours. Add black pepper sauce and boil till the soup is thick.

3. Make 10 potato balls and roll in salt. Deep-fry them in 180°C cooking oil till golden yellow. Add the lamb soup and boil the potato balls to be well done.

4. Place the lamb chops on the plate and place the potato balls around them.

NOTES

Braised Lamb Chops with Potatoes is made of specially-selective black goat meat of Lezhi. With the western cooking methods, the dish is reddish brown, and the chops are tender, fresh, and salty.

山椒熘羊肝
65 STIR-FRIED LAMB LIVER WITH MOUNTAIN PEPPERS

食材配方

黑山羊肝200克　青笋100克
水发木耳50克　野山椒10克
泡辣椒20克　　泡仔姜10克
蒜片10克　　　马耳朵葱10克
食盐6克　　　　料酒6毫升
酱油5毫升　　　醋2毫升
白糖1克　　　　水淀粉50克
鲜汤10毫升　　食用油50毫升

制作工艺

1. 黑山羊肝切成薄片，加入食盐、料酒、水淀粉上浆；青笋切成菱形片；水发木耳切小块；野山椒、泡辣椒切马耳朵形；泡仔姜切指甲片。

2. 将食盐、料酒、酱油、醋、白糖、水淀粉、鲜汤调成芡汁。

3. 锅置火上，放油烧至180℃，放入羊肝炒散至熟，入野山椒、泡辣椒、泡仔姜、蒜片、葱、青笋片、木耳炒香至熟，倒入芡汁，收汁亮油后起锅装盘即成。

评鉴

羊肝富含维生素A和磷，有益肝、明目的作用。将羊肝搭配野山椒、泡姜、泡辣椒一同烹炒，能除膻增香，成菜味道酸辣咸鲜，羊肝质地细嫩。

INGREDIENTS

200g black goat liver; 100g stem lettuce; 50ml water-soaked black fungus; 10g mountain peppers; 20g pickled chilies; 10g pickled tender ginger; 10g garlic, sliced; 10g green onions, segmented diagonally; 6g salt; 6ml Shaoxing cooking wine; 5ml soy sauce; 2ml vinegar; 1g sugar; 50g batter; 10ml stock; 50ml cooking oil.

PREPARATION

1. Cut the black lamb liver into slices, coating it with salt, Shaoxing cooking wine, and batter. Cut the stem lettuce into diamond slices. Slice the water-soaked black fungus. Slice the mountain peppers and pickled chilies diagonally. Cut the pickled tender ginger into the size of a fingernail.

2. Make sauce with salt, Shaoxing cooking wine, soy sauce, vinegar, sugar, batter and stock.

3. Heat up the oil to 180°C in a wok and add the black lamb liver. Fry until it is cooked through, then add the mountain peppers, pickled chilies, pickled tender ginger, garlic, spring onion, stem lettuce, and black fungus. Add and thicken the sauce and transfer to the serving dish.

NOTES

Lamb liver is rich in vitamin A and phosphorus, which benefit people's liver and eyes. By frying the lamb liver with mountain peppers, pickled ginger and pickled chilies, the lamb smell can be eliminated and the delicious aroma can be strengthened. The dish has sour, spicy, salty and savory flavor and tastes fresh and tender.

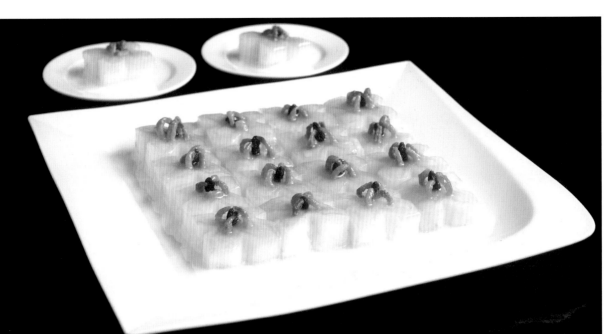

66 金钩瓜烹 STEAMED WHITE GOURD WITH DRIED SHRIMPS

食材配方

冬瓜1 500克
枸杞12颗
猪油50克
姜片15克
胡椒粉0.5克
鸡油20克
金钩100克
鲜汤200毫升
葱节50克
食盐5克
水淀粉10克

INGREDIENTS

1,500g white gourd; 100g dried sea shrimps; 12 wolfberries; 200ml stock; 50g lard; 50g spring onions, segmented; 15g ginger, sliced; 5g salt; 0.5g pepper; 10g batter; 20g chicken fat

制作工艺

1 冬瓜切成约5厘米见方、2.5厘米厚的正方块并雕成花，在每块瓜方的表面镶嵌上金钩和干枸杞，放入方盘，加入鲜汤、食盐，入笼蒸10分钟至冬瓜成熟，出笼后装盘。

2 锅置火上，放猪油烧至120℃，放入姜片、葱节炒香，倒入蒸瓜方的汤，入食盐、胡椒粉、鸡油、水淀粉，调味收汁成清二流芡，淋在瓜方上即成。

PREPARATION

1. Cut the white gourd into about 5cm long and 2.5cm thick cubic chunks, and carve them into the shape of a flower. Inlay the chunks with dried sea shrimps and dried wolfberries. Put the inlaid chunks in the plate and add stock and salt. Put the plate in a steamer and steam for 10 minutes till the white gourd is cooked through. Transfer to the serving dish.

2. Heat the oil to 120°C and add ginger slices and spring onions. Stir till aromatic and add the steamed chunks' juice, add salt, peppers, chicken oil and batter. Stir fry the mixed seasoning till it is reduced to semifluid. Pour the sauce over the chunks.

评鉴

此菜为夏季时令菜品，色泽鲜艳，味道咸鲜醇厚，清爽可口。

NOTES

The dish is a kind of seasonal summer dish. It is bright in color and salty, savory and refreshing in flavor.

67 资阳手撕烤兔
ZIYANG HAND SHREDDED ROAST RABBIT

食材配方

鲜兔1只（约1000克）　折耳根100克
葱花20克　　　　　　洋葱丁30克
食盐10克　　　　　　料酒50毫升
酱油20毫升　　　　　白糖10克
花椒5克　　　　　　　干辣椒15克
姜片10克　　　　　　孜然10克
辣椒油50毫升

制作工艺

1 鲜兔用清水浸泡30分钟，去除血水后捞出，沥干水分，加入食盐、料酒、酱油、白糖、花椒、干辣椒、姜片拌匀，腌制4~5小时。

2 将腌制好的兔肉撒上孜然，放进烤炉，用小火慢烤40分钟后取出，撕成丝。

3 将兔肉丝、折耳根、葱花、洋葱丁、辣椒油拌匀，装盘即成。

评鉴

资阳手撕烤兔是资阳市的代表性名菜，选料考究，制法独特，口味香醇，肉厚处醇香绵软，肉薄处酥香脆爽，细细咀嚼，口齿留香。

INGREDIENTS

1 rabbit (about 1,000g); 100g fish mint; 20g spring onions, finely chopped; 30g onions, diced; 10g salt; 50ml Shaoxing cooking wine; 20ml soy sauce; 10g sugar; 5g Sichuan pepper; 15g dry chilies; 10g ginger, sliced; 10g cumin; 50ml chili oil

PREPARATION

1. Soak the rabbit in water for 30 minutes before blanching to remove extra blood, drain, and marinate with salt, Shaoxing cooking wine, soy sauce, sugar, Sichuan pepper, dried chilies and ginger for 4 to 5 hours.

2. Add cumin to the marinated rabbit and roast it with low heat for 40 minutes, and shred with hand.

3. Combine the shredded rabbit, fish mint, spring onions, onions and chili oil, stir well and transfer to a serving dish.

NOTES

Hand-shredded Roast Rabbit, a typical dish in Ziyang, uses quality ingredients with peculiar cooking methods. The dish is worthy of chewing, leaving aromas in the mouth and between the teeth.

食材配方

紫薯150克　　　山药100克
老南瓜150克　　玉米棒100克
火腿酱40克　　 食盐5克
白糖20克　　　 清水200毫升
食用油500毫升（约耗30毫升）

制作工艺

1 紫薯、山药、南瓜、玉米棒分别切成长约8厘米大的块，放入高压锅内，加入食用油、少量清水，加热压2分钟。

2 火腿酱、食盐、白糖、清水入锅烧沸制成酱汁。

3 将紫薯、山药、南瓜、玉米棒摆放在煲仔内，加入酱汁，煲1分钟即成。

评鉴

紫薯、山药、南瓜、玉米的共性是脂肪含量低，富含维生素、纤维素等，有一定的促进肠道蠕动、降低血压的作用。此菜以它们为食材制成，味道咸鲜，略带回甜，酱香浓郁。

INGREDIENTS

150g purple sweet potatoes; 100g Chinese yams; 150g pumpkins; 100g corns on the crop; 40g ham sauce; 5g salt; 20g sugar; 200ml water; 500ml cooking oil (about 30ml to be consumed).

PREPARATION

1. Cut purple sweet potatoes, Chinese yams, pumpkins, and corns into about 8cm chunks and transfer to a pressure cooker. Add cooking oil, and a small amount of water, and heat for 2 minutes.

2. Make the sauce by boiling the ham sauce, salt, sugar and water.

3. Place the purple sweet potatoes, Chinese yams, pumpkins, and corns in the wok, add the sauce, and simmer for 1 minute.

NOTES

Purple sweet potatoes, Chinese yams, pumpkins and corns are of low fat, rich vitamins and rich cellulose, which have certain effects of promoting intestinal peristalsis and lowering blood pressure. It tastes savory, slightly sweet, and has a great smell of ham sauce.

68 火腿酱四宝
FOUR TREASURES IN HAM SAUCE

69 东安云白豆腐

食材配方

烟熏豆腐400克　　肥肉丝20克

蒜10克　　　　　鸡精2克

食盐2克　　　　　葱白20克

水淀粉30克　　　 鲜汤600毫升

化猪油200毫升

制作工艺

1. 烟熏豆腐去皮，切成长约8厘米的细丝，入沸水锅焯水2~3次，再入冷水浸泡1小时。

2. 肥肉切成长约8厘米的细丝；葱白切成细丝，入清水浸泡；蒜切成细丝。

3. 锅置火上，放猪化油烧至80℃，放入蒜丝、肥肉丝炒断生，掺入鲜汤，放入食盐、鸡精、豆腐丝烧沸，入水淀粉勾芡，收汁亮油后出锅装盘，撒上葱丝即成。

评鉴

云白豆腐丝是川菜传统名品，清代光绪年间，四川总督府专用厨师（别号"斋二师"）为四川总督刘秉璋创制，据传因初创时晴空万里，天空飘来朵朵白云，总督因此赐名"云白豆腐丝"。此菜色白如云，质地嫩滑，咸鲜清香，油而不腻。

INGREDIENTS

400g smoked doufu; 20g fatty pork; 10g garlic; 2g chicken essence; 2g salt; 20g spring onion white; 30g batter; 600ml stock; 200ml lard

PREPARATION

1. Peel the smoked doufu and slice into about 8cm slivers. Blanch in boiling water 2 to 3 times and soak in cold water for 1 hour.

2. Slice the pork fat into 8cm slivers. Slice the spring onion white into slivers and soak in water. Slice the garlic into slivers.

3. Heat the lard in the wok to 80℃. Stir fry garlic slices and fat slivers in the wok. Add the stock, salt, chicken essence, and doufu slivers and heat till the stock is boiling. Add the batter and reduce the stock before the dish is lustrous. Serve at once with spring onion slivers.

NOTES

White Cloud Doufu slivers is a traditional Sichuan dish. During the reign of Emperor Guangxu in the Qing Dynasty, the chef for Sichuan Governor Liu Bingzhang invented the dish. It was recorded that, on the day when the dish was served, the sky was clear with white clouds, so the governor called this dish Cloud White Doufu Slivers. The doufu is white as cloud, and tastes salty, tender, fatty but not greasy.

70 麻香汩水老豆腐

SICHUAN PEPPER FLAVOR HARD DOUFU

食材配方

泹水老豆腐500克	红小米椒30克
青椒100克	洋葱80克
香菜60克	大葱50克
姜50克	蒜100克
花椒油50毫升	蚝油40毫升
鸡粉20克	味精2克
美极鲜酱油20毫升	芥末20克
辣鲜露10克	白糖10克
青花椒100克	食盐10克
葱花5克	鲜汤300毫升
菜籽油300毫升	

制作工艺

1. 泹水老豆腐切成厚块；红小米椒、青椒、洋葱、香菜、大葱切碎。

2. 锅置火上，入红小米椒、青椒、洋葱、香菜、大葱、姜、蒜、花椒油、蚝油、鸡粉、味精、美极鲜酱油、芥末、辣鲜露、白糖、食盐、鲜汤熬制成味汁。

3. 锅置火上，放菜籽油烧至150℃，放入姜、葱、蒜爆香，入味汁、老豆腐块烧沸3分钟，出锅装盘。

4. 锅置火上，放菜籽油烧至100℃，放入青花椒炒香，连油一起淋在豆腐上，撒上葱花即成。

评鉴

资阳境内许多家庭都有自制泹水豆腐的习惯，当地厨师在传承川味的同时大胆创新，此菜就是用当地产的青花椒为主要调料创制的豆腐新菜，集鲜、麻、香、烫、嫩为一体，给人不一样的味觉冲击。

INGREDIENTS

500g hard Doufu; 30g red bird's eye chilies; 100g peppers; 80g onions; 60g green corianders; 50g spring onions; 50g ginger; 100g garlic; 50ml Sichuan pepper oil; 40ml oyster sauce; 20g chicken essence; 2g MSG; 20ml Maggi soy sauce; 20g mustard; 10g spicy fresh dew seasoning; 10g sugar; 100g Sichuan green pepper; 10g salt; 5g chives, finely chopped; 300ml stock; 300ml rapeseed oil

PREPARATION

1. Cut the hard Doufu into thick chunks; finely chop the red bird's eye chilies, green peppers, onions, corianders and spring onions.

2. Heat the wok and add red bird's eye chilies, green peppers, onions, green corianders, spring onions, ginger, garlic, Sichuan pepper oil, oyster sauce, chicken essence, MSG, Maggi soy sauce; mustard, spicy fresh dew seasoning, sugar, salt and stock. Boil them into seasoning sauce.

3. Heat the rapeseed oil to 150℃ and put ginger, spring onions and garlic in, stir fry till aromatic and add the prepared seasoning sauce and the hard Doufu, boil them for 3 minutes and transfer to the serving dish.

4. Heat the rapeseed oil to 100℃ and put Sichuan green pepper in, stir fry till aromatic. Pour the Sichuan green peppers with oil over the Doufu and sprinkle with chopped chives.

NOTES

Many families in Ziyang City of Sichuan have the tradition to make Doufu. The local cooks create new dishes with boldness and creativity while they inherit the traditional Sichuan cuisines. This dish is a new cuisine of Doufu and its major seasoning is the local Sichuan green pepper. It is a balance of savory, hot, dainty and tender tastes.

食材配方

野猪肉700克	嫩蛋100克
大蒜50克	青笋100克
小米辣10克	青尖椒20克
香菜10克	香葱20克
八角3克	桂皮3克
当归3克	香叶2克
小茴1克	山柰2克
干辣椒8克	花椒5克
姜片10克	秘制复合油100毫升
临江寺豆瓣50克	啤酒50毫升
味精2克	鲜汤500毫升

制作工艺

1. 野猪肉斩成块；青笋切成滚料块，煮熟；青尖椒、小米辣分别斜切成节；嫩蛋切块，入油锅略炸。

2. 锅置火上，放入秘制复合油烧至160℃，放入野猪肉煸炒至香，入临江寺豆瓣、干辣椒、花椒、姜片、大蒜、八角、桂皮、当归、香叶、小茴、山柰炒香，加入啤酒、鲜汤烧沸，倒入高压锅压至野猪肉软熟，倒入锅中，去掉料渣，放入青尖椒、小米辣，收汁浓稠后放入味精调味。

3. 将嫩蛋、青笋块放入吊锅中垫底，放上野猪肉，点缀香菜，上桌时点燃吊锅下面的酒精炉加热食用。

评鉴

这里选用人工饲养的特种野猪肉，味道鲜美，蛋白质、亚油酸及亚麻酸含量高，胆固醇含量低、脂肪含量少。此菜以它为原料制成，味道咸鲜香辣，皮软糯、有嚼头，配上软嫩的鸡蛋，更形成了口感上的互补。

INGREDIENTS

700 boar meat; 100g egg, steamed; 50g garlic; 100g stem lettuce; 10g bird's eye chilies; 20g green chili peppers; 10g coriander; 20g chives; 3g star anise; 3g cinnamon; 3g Chinese angelica; 2g myrcia; 1g fennel; 2g sand ginger; 8g dried chilies; 5g Sichuan pepper; 10g ginger, sliced; 100ml homemade cooking oil; 50g Linjiangsi chili bean paste; 50ml beer; 2g MSG; 500ml stock

PREPARATION

1. Chunk the boar meat and stem lettuce, and boil to be well done. Cut the green chili peppers and bird's eye chilies into sections. Slice the steamed eggs into chunks and deep-dry.

2. Heat the homemade cooking oil to 160°C in the wok. Fry the boar meat and blend in Linjiangsi chili bean paste, dried chili pepper, Sichuan pepper, ginger, garlic, star anise, cinnamon, Chinese angelica, myrcia, fennel, sand ginger and fry. Mix beer with stock, boil and pour into the pressure cooker with the boar meat. Heat the cooker till the meat is soft and transfer to the wok and remove condiment residues. Add some green chili peppers and bird's eye chilies and heat till the soup is thick. Add MSG.

3. Place steamed eggs and sterm lettuce at the bottom of the hanging wok and place the meat on them. Garnish with coriander. Light the alcohol heater under the hanging wok when serving.

NOTES

The boar is raised by locals and tastes delicate and fresh. This meat is rich in protein, linoleic acid and linolenic acid and has only a small amount of cholesterol and fat. This dish is salty, fresh and sticky and tastes greater coupled with the tender steamed eggs.

71 风味野猪肉
SAVORY BOAR MEAT

72 苕皮回锅肉

TWICE-COOKED PORK WITH SWEET POTATO NOODLES

食材配方

猪带皮二刀肉350克　苕粉皮150克
蒜苗40克　　　　　　临江寺豆瓣30克
甜面酱10克　　　　　白糖3克
酱油3毫升　　　　　　味精2克
食用油100毫升

制作工艺

1 将猪带皮二刀肉放入锅中煮熟，捞出凉冷，切成长约6厘米、宽约4厘米、厚约0.15厘米的片。

2 苕粉皮切成菱形片；蒜苗切成马耳朵形。

3 锅置火上，放油烧至150℃，放入肉片炒成"灯盏窝"形，入临江寺豆瓣炒香、油呈红色，入甜面酱、白糖、苕粉皮、酱油炒匀，再放入蒜苗、味精炒熟，起锅装盘即成。

评鉴

此菜是在川菜传统名菜回锅肉的基础上加入安岳红薯粉皮制作而成，色泽红亮，咸鲜微辣，回味略甜，肉片滋润，苕皮软糯。

INGREDIENTS

350g pork rump with skin attached; 150g sweet potato noodles; 40g green garlic; 30g Linjiangsi chili bean paste; 10g fermented flour paste; 3g sugar; 3ml soy sauce; 2g MSG; 100ml cooking oil

PREPARATION

1. Boil the pork until cooked through, remove and cool, then cut into slices about 6cm long, 4cm wide and 0.2cm thick.

2. Cut the sweet potato noodles; cut the green garlic into horse-ear shape slices.

3. Heat oil in a wok to 150℃, add the pork to stir-fry till the pork slices slightly curl up. Blend in Linjiangsi chili bean paste to stir-fry to bring out aroma, add fermented flour paste, sugar, sweet potato noodles and soy sauce and mix well and continue to stir-fry till aromatic. Add the green garlic and MSG to stir-fry till al-dente. Remove from the heat and transfer to a serving dish.

NOTES

This dish is localized by adding Anyue sweet potato noodles to the traditional Sichuan famous dish Twice-cooked Pork. This dish is reddish brown and lustrous, and has a salty, sweet and slightly hot taste, delicate and juicy meat, soft and glutinous sweet potato noodles.

73 鸡杂粉条
BRAISED CHICKEN OFFAL WITH VERMICELLI

食材配方

红薯粉条150克　鸡杂100克
临江寺豆瓣25克　食盐5克
味精2克　　　　大葱15克
香葱5克　　　　醋20毫升
鲜汤500毫升　　食用油100毫升

制作工艺

1. 红薯粉条用热水泡软；鸡杂切片，大葱切粒；香葱切成葱花。

2. 锅置火上，放油烧至160℃，放入鸡杂炒熟，入临江寺豆瓣、大葱炒香，掺入鲜汤烧沸出香味，加入粉条、食盐、味精、醋烧沸，起锅装盘，放入葱花成菜。

评鉴

此菜选用安岳特产的红薯粉条制作成菜，酸辣爽口，粉条劲道滑爽，形整不断，色泽鲜亮，鸡杂香脆可口。

INGREDIENTS

150g sweet potato vermicelli; 100g chicken offal; 25g Linjiangsi chili bean paste; 5g salt; 2g MSG; 15g spring onions; 5g chives; 20ml vinegar; 500ml stock; 100ml cooking oil

PREPARATION

1. Soften the sweet potato vermicelli in hot water. Slice the offal and finely chop the spring onions and chives.

2. Heat the cooking oil in the wok to 160 °C. Add the chicken offal and fry. Blend in the Linjiangsi chili bean paste and spring onions and fry. Add the stock, vermicelli, salt, MSG, and vinegar and bring to a boil. Serve at once with chives.

NOTES

This dish is made from the sweet potato vermicelli of Anyue. It is delicious and refreshing. The vermicelli is smooth, the color is lustrous, and the chicken offal is crispy and delicious.

74 口袋豆腐
SACK DOUFU

沥水豆腐600克　　碎肉250克
食盐6克　　　　　味精2克
姜汁10毫升　　　　料酒5毫升
水淀粉10克　　　　西兰花300克
韭菜20克　　　　　紫甘蓝150克
鸡汁50毫升　　　　鲜汤1 000毫升

1 碎肉中加入食盐、姜汁、料酒、水淀粉拌匀成馅。

2 将沥水豆腐的内部掏空，填入肉馅，入锅炸至色泽金黄后捞出，用韭黄叶捆扎成口袋形，再入鲜汤煮熟。

3 将紫甘蓝削成碗状，放入煮熟的西兰花和口袋豆腐。

4 将煮豆腐的原汤倒入锅中，加入鸡汁、食盐、味精、水淀粉收汁浓稠，浇淋在豆腐上即成。

由于乐至县蟠龙湖的水质很好，用这里的水制作出的豆腐品质更佳，口感嫩滑，回味微甜。此菜因将豆腐制成口袋形而得名，色泽淡黄，味道咸鲜。

INGREDIENTS

600g doufu; 250g pork mince; 6g salt; 2g MSG; 10ml ginger juice; 5ml Shaoxing cooking wine; 10g batter; 300g broccoli; 20g Chinese leek; 150g red cabbage; 50ml chicken broth; 1,000ml stock

PREPARATION

1. Mix minced pork well with salt, ginger juice, Shaoxing cooking wine and batter to make the stuffing.

2. Empty the doufu and fill in with the stuffing. Heat oil in a wok, deep-fry the doufu until golden brown, remove and tie the doufu with the leeks as a sack, then boil in the stock until cooked through.

3. Cut the red cabbage as a bowl, put the cooked broccoli and the sack doufu in.

4. Pour the boiled doufu stock into a wok, add chicken broth, salt, MSG and batter and simmer until thicken, pour it over the doufu then serve.

NOTES

Lezhi County has high quality water source because of the Panlong Lake. Doufu made by the water from here has greater quality, tender and slightly sweet taste. Doufu in this dish shapes like a sack, hence its name. This dish has a golden-brown color, a salty and savory taste.

叁 面点小吃
Episode Three: Snacks

食材配方

面粉500克　　黄油200毫升
猪肉200克　　莲蓉200克
花生米50克　　白糖50克

制作工艺

1. 面粉与黄油和匀，调制成油酥面团
2. 花生米烤香酥，去皮后压碎，加入莲蓉、白糖拌匀，制成莲蓉花生馅
3. 油酥面团包入莲蓉花生馅，入模具制成花生形状，入烤箱烤熟，取出装盘即成。

评鉴

花生也称长生果，被视为吉祥喜果，有生生不息的寓意。此菜形象逼真，口味香甜，具有浓郁的地方特色。

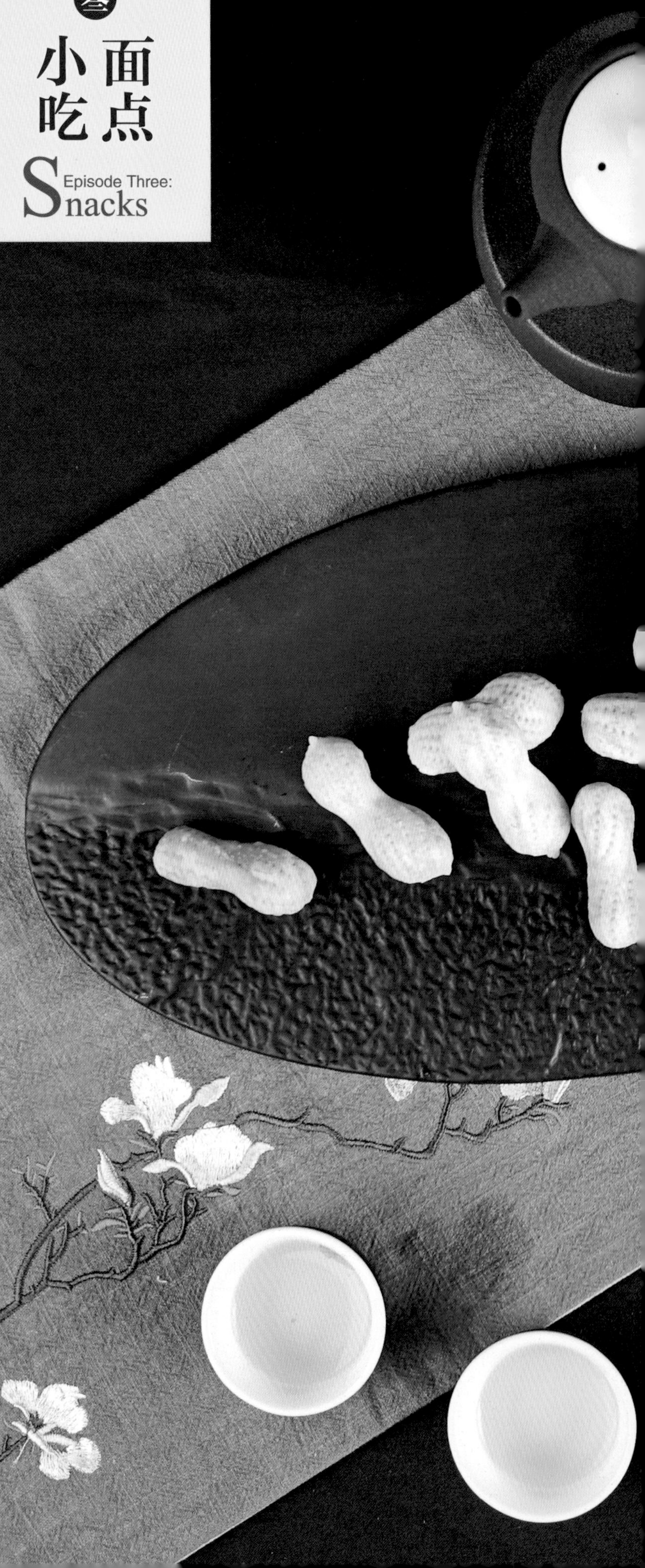

01 长生仙果
LONGEVITY FRUITS

INGREDIENTS
500g wheat flour; 200ml butter; 200g pork, 200g lotus seed paste; 50g peanuts; 50g sugar

PREPARATION
1. Mix the wheat flour with butter, and knead to make the oil dough.

2. Roast the peanuts, remove the skin and crush. Mix well with lotus seed paste and sugar to make the filling.

3. Wrap up the filling with the dough, and mould into a shape of peanuts. Place the peanuts into the oven, and bake till cooked through.

NOTES
Peanuts, also called longevity fruit, are regarded as a symbol of auspiciousness and reproduction. This vivid dish is a local specialty with sweet, luscious tastes.

02 坛子肉粑粑

BABA CAKES WITH TANZI PORK

食材配方

坛子肉150克　面粉500克
生粉100克　汤圆粉150克
食盐2克　味精2克
葱花200克　食用油30毫升
清水750毫升

制作工艺

1. 将面粉、生粉、汤圆粉入盆，加入清水调匀，再放入食盐、味精、葱花搅匀成面糊；坛子肉切成薄片。

2. 平底锅置火上，放油烧至120℃，放入面糊，摊为圆饼状，再放入1片坛子肉，煎至两面呈金黄色且成熟后出锅装盘即成。

评鉴

"粑粑"是四川民间方言，泛指饼类食物，各地制法有一定差异。此菜融入了资阳本土特产坛子肉制作而成，口感软糯，坛子肉味香浓郁。

INGREDIENTS

150g Tanzi Pork; 500g wheat flour; 100g cornstarch; 150g glutinous rice flour; 2g salt; 2g MSG; 200g spring onions, finely chopped; 30ml cooking oil; 750ml water

PREPARATION

1. Put flour, cornstarch and glutinous rice flour into a pot and add some water to mix them well. Then add salt, MSG, spring onions, stir all the ingredients well to a paste. Cut the Tanzi pork into thin pieces.

2. Heat the oil to 120℃ in a pan, put the paste in and make it round, then add one piece of Tanzi pork and fry till both sides are golden and transfer to the serving dish.

NOTES

"Baba" belongs to the vocabulary of the folk dialect in Sichuan, which generally refers to pastry, and different places have different cooking methods. This dish is made with the "Tanzi" pork, which is the local specialty in Ziyang city of Sichuan and it refers to meat in a clay jar. It tastes soft and glutinous with a strong delicious aroma of Tanzi Pork.

陈氏凉粉

03 CHEN'S PEA JELLY

食材配方

豌豆300克　　　清水1 000毫升
红油辣椒150克　花椒粉15克
蒜泥20克　　　葱花50克
熟芝麻10克　　酥花生碎30克
食盐3克　　　　豆豉20克
酱油10毫升　　水淀粉30克

制作工艺

1 豌豆用冷水浸泡10小时，磨成汁，过滤沉淀，将沉淀的精粉与一部分粉汁取出，入锅中加水搅至糊状，盛入容器内冷却后制成凉粉，再切成长粗丝。

2 豆豉剁成细末，入锅中炒散、出香，加清水煮5～6分钟，加入食盐、酱油，用水淀粉勾成二流芡，制成豆豉卤汁。

3 凉粉装入碗中，依次加入豆豉卤汁、红油辣椒、花椒粉、蒜泥、熟芝麻、酥花生碎、葱花拌匀即成。

评鉴

陈氏凉粉是陈毅元帅的侄孙女陈丽芸女士所制作的陈家小吃代表性品种。该菜品选用自制的豌豆黄凉粉为食材，温润如玉，色泽红亮，味道咸鲜麻辣，豉香浓郁。

INGREDIENTS

300g peas; 1,000ml water; 150g chili oil; 15g ground Sichuan pepper; 20g garlic; 50g spring onions, finely chopped; 10g roasted sesames; 30g peanuts, roasted and crushed; 3g salt; 20g fermented soybeans; 10ml soy sauce; 30g batter

PREPARATION

1. Soak the peas in water for 10 hours, grind and leave to settle. Transfer the sediments to a pot, add water, and stir till a paste has formed. Transfer the paste to a container, and leave to cool. Cut the jelly into thick slivers.

2. Finely chop the fermented soybeans, and stir fry in a wok till aromatic. Add water, and leave to boil for 5 to 6 minutes. Blend in the salt, soy sauce and batter to make the seasoning sauce.

3. Place the jelly in a bowl, pour over the seasoning sauce, and add the chili oil, Sichuan pepper, garlic, sesames, peanuts and spring onions.

NOTES

Chen's Jelly is one of the typical snacks made by Chen Liyun, grand-niece of Marshal Chen Yi. This dish features hand-made jelly, reddish color, and salty, spicy tastes with a strong fermented soybean aroma.

食材配方

豌豆300克　　清水1 000毫升
红油辣椒150克　花椒粉15克
姜汁20毫升　　蒜泥20克
葱花50克　　　熟芝麻10克
酥花生碎30克　食盐2克
水淀粉30克

制作工艺

1 豌豆用冷水泡10个小时，磨成汁，过滤沉淀，将沉淀的精粉和一部分水汁取出，入锅中加清水搅至糊状，盛入容器中冷却后成凉粉。

2 食盐、水淀粉入锅，调成咸酱汁。

3 凉粉切长粗丝，装入碗中，依次加入咸酱汁、红油辣椒、花椒粉、姜汁、蒜泥、熟芝麻、酥花生碎、葱花拌匀即成。

评鉴

周礼伤心凉粉始于20世纪初，由文江源在传统黄凉粉风味基础上改进而成。20世纪80年代，文江源将其制作技艺传授给女婿姚长明，品质和滋味更胜一筹。此小吃色泽鲜亮，口感爽滑，味道咸鲜香辣。

INGREDIENTS

300g peas; 1,000ml water; 150g chili oil; 15g ground Sichuan pepper; 20ml ginger juice; 20g garlic, crushed; 50g spring onions, finely chopped; 10g roasted sesames; 30g crispy peanuts, crushed; 2g salt; 30g batter

PREPARATION

1. Soak the peas in water for 10 hours, grind and leave to settle. Transfer the sediments to a pot, add water, and stir till a paste has formed. Transfer the paste to a container, and leave to cool.

2. Add salt and batter, mix them well to make the salty sauce.

3. Cut the jelly into thick slivers and remove to a bowl. Add the salty sauce, chili oil, ground Sichuan pepper, ginger juice, garlic, sesames, peanuts and spring onions and mix them well.

NOTES

This dish originated from the beginning of 20th century. It was improved by Wen Jiangyuan on the basis of traditional yellow jelly. In 1,980s, Wen passed on the skill of making this dish to his son-in-law, Yao Changming, who further improved the quality and flavor of the dish. Zhouli spicy pea jelly is bright in color, smooth and crispy in taste and salty, delicious and spicy in flavor.

05 红油米卷
RICE ROLLS IN CHILI OIL

食材配方

安岳米卷250克
老抽5毫升
味精1克
白糖10克
鲜汤10毫升
红油80毫升
生抽20毫升
香油5毫升
葱花15克

制作工艺

1 米卷切成长约1.5厘米的节，展开后装入盘中。

2 将红油、老抽、生抽、味精、香油、白糖、鲜汤调制成红油味汁，淋在米卷上，撒上葱花即成。

评鉴

安岳米卷为安岳县特色食材，历史悠久，做工精细，具有色泽光亮、薄如纸张、口感鲜美、质地嫩滑等特点。此菜用红油味汁调味，味道咸鲜香辣，略带回甜，老幼皆宜。

INGREDIENTS

250g Anyue rice rolls; 80ml chili oil; 5ml dark soy sauce; 20ml light soy sauce; 1g MSG; 5ml sesame oil; 10g sugar; 15g spring onions, finely chopped; 10ml stock

PREPARATION

1. Cut the rice rolls into about 1.5cm segments, tear apart and put on the plate.

2. Make the sauce of chili oil, dark soy sauce, light soy sauce, MSG, sesame oil, sugar and stock. Sprinkle the sauce and spring onions on the rice rolls.

NOTES

Anyue Rice Rolls in Chili Oil is a special dish in Anyue County, Ziyang City. This dish, with a long history, has bright color and great flavor, and the rolls are as thin as pieces of paper. This dish is seasoned with chili oil and tastes salty, spicy and a little sweet and is appreciated by both the young and the old.

食材配方

千层油皮12张　柠檬1个
猪板油50毫升　白糖50克
蜜瓜条30克　　酥花生30克
鸡蛋黄1个　　　熟芝麻15克
熟面粉40克

制作工艺

1. 猪板油、蜜瓜条、酥花生分别剁细；柠檬取汁。

2. 将猪板油、柠檬汁、白糖、蜜瓜条、酥花生、熟芝麻、熟面粉拌匀成柠檬馅。

3. 将柠檬馅放在千层油皮上，对折成半圆形，刷上蛋黄液，装入烤盘，入烤箱烤制20分钟，取出后装盘即成。

评鉴

此面点酥香化渣，酥层的层次分明，柠檬香味突出。

INGREDIENTS

12 lard short dough; 1 lemon; 50ml lard; 50g sugar; 30g winter melon candy bars; 30g crispy peanuts; 1 egg yolk; 15g roasted sesames; 40g cooked wheat flour

PREPARATION

1. Chop lard, winter melon candy bars and crispy peanuts. squeeze out lemon juice.

2. Blend lard, lemon juice, sugar, winter melon candy bars, sugar, peanut, sesames and cooked wheat flour to make lemon stuffing.

3. Roll and flat the dough, put the lemon stuffing on it, fold into half, brush with whisked egg yolk. Put on the baking plate and bake for 20 minutes in preheated oven. Then take out and place them on a serving plate.

NOTES

This pastry cake is super crispy and has lemon fragrance. It has nice multi-layered look.

06 柠檬酥
LEMON PIE

07 王辣面

WANG'S SPICY NOODLES

食材配方

湿面条150克	猪肉100克
芝麻油5毫升	食用油20毫升
姜米5克	蒜米10克
泡辣椒末10克	青尖椒10克
小米辣5克	食盐10克
鸡精2克	

制作工艺

1 猪肉切成丝；青尖椒、小米辣切成长约1.5厘米的节。

2 锅置火上，放油烧至160℃，放入肉丝炒干水分，入姜米、蒜米炒香，再入泡辣椒末、青尖椒、小米辣炒熟，最后入芝麻油、食盐、鸡精制成面臊。

3 锅置火上，放水烧沸，入湿面条煮熟，捞入面碗，放入面臊即成。

评鉴

王辣面是资阳市名小吃，油色透亮，干香味浓，辣鲜皆备。其面臊较多，除炒肉丝外，还有烧肉等。

INGREDIENTS

150g noodles; 100g pork; 5ml sesame oil; 20ml cooking oil; 5g ginger, finely chopped; 10g garlic, finely chopped; 10g pickled chilies, finely chopped; 10g green chilies; 5g bird's eye chilies; 10g salt; 2g chicken essence

PREPARATION

1. Cut the pork into shreds. Cut the green chilies and bird's eye chilies into about 1.5cm segments.

2. Heat the oil to 160°C in a wok, put the shredded pork and fry till they are almost dry. Add the ginger and garlic to fry till aromatic, then add the pickled chilies, green chilies and bird's eye chilies to fry. Finally, add sesame oil, salt and chicken essence to make the noodle topping.

3. Heat the pot and add water till it is boiling, then add noodles till well cooked. Remove the noodles and put them in a bowl, add the noodle topping.

NOTES

Wang's Spicy Noodles is a famous snack in Ziyang city of Sichuan, which tastes spicy with a strong delicious aroma in flavor. It has lots of noodle topping, including stir-fried shredded pork, braised pork, etc.

08 乐至麻饼
SESAMES PASTRY LEZHI

食材配方

面粉100克
花生100克
葡萄干100克
橘红100克
植物油100毫升
瓜子仁100克
芝麻100克
冬瓜糖100克
白糖100克

制作工艺

1 将瓜子仁、花生、葡萄干、冬瓜糖、橘红拌匀，揉捏成馅心。

2 面粉、植物油、白糖和成面团，下剂，按压成面皮，放入馅心包裹成团，擀成圆饼状，撒上芝麻，放入烤箱烘烤30分钟至色金黄、皮酥香后取出装盘即成。

评鉴

乐至麻饼是当地传统糕点，采用纯手工操作，色泽黄而不焦，皮酥心脆，味道香甜。

INGREDIENTS

100g wheat flour; 100g sunflower seeds; 100g peanuts; 100g sesames; 100g raisins; 100g winter melon candy bars; 100g dried orange peel; 100g sugar; 100ml vegetable oil

PREPARATION

1. Mix sunflower seeds, peanuts, raisins, winter-melon candy bars and dried orange peel and blend well, knead the ingredients to make the filling.

2. Mix the flour with vegetable oil and sugar and knead into a dough, roll the dough into a rope and segment, press each segment into flat and round wrappers. Place the filling on it and wrap up. Press it slightly into a round pastry cake and sprinkle with sesames on both sides. Bake in the oven for 30 minutes until golden brown and crispy, remove and transfer to a serving plate.

NOTES

As a local traditional cake, Lezhi Sesames Pastry has golden brown color, crispy skin but crunch inside, slightly sweet and aromatic taste.

食材配方

面粉500克　　　牛瘦肉150克
鸡蛋液50克　　 食盐5克
葱花10克　　　 淀粉10克
花椒粉0.1克　　辣椒粉0.1克
料酒5毫升　　　猪油150克
食用油100毫升　清水100毫升

INGREDIENTS

500g wheat flour; 150g lean beef; 50g beaten egg; 5g salt; 10g spring onions; 10g starch; 0.1g Sichuan pepper; 0.1g ground chilies; 5ml Shaoxing cooking wine; 150g lard; 100ml cooking oil; 100ml water

制作工艺

1. 牛瘦肉剁碎，加入食盐、料酒、葱花、花椒粉、辣椒粉、鸡蛋液、淀粉拌匀成牛肉馅。
2. 一半面粉加入清水、食盐、猪油50克调成油水面团；另一半面粉加入猪油100克调成油酥面团。
3. 将油水面团包入油酥面团擀成薄皮，卷成长条，切成小节擀成薄皮，再包入牛肉馅，制成钟焦粑初坯。
4. 平底锅置火上，入油烧热，放入钟焦粑初坯煎至两面金黄色时捞出，待油温回升后再将初坯放入炸熟，出锅装盘即成。

PREPARATION

1. Finely chop the beef, and blend well with the salt, Shaoxing cooking wine, spring onions, Sichuan pepper, ground chilies, beaten egg and starch to make the filling.
2. Combine half of the wheat flour with water, salt and lard (50g), and knead to make the water-lard dough. Combine the remaining wheat flour with lard (100g), and knead to make the lard dough.
3. Wrap up the water-lard dough in the lard dough, and roll flat with a rolling pin. Roll up the sheet, cut into segments, and roll with a rolling pin into thin wrappers. Seal up the beef fillings with the wrappers.
4. Heat oil in a pan over the heat, and fry both sides of the beef dough till golden brown. Remove from the pan, reheat the oil and then deep fry the beef pie. Transfer to a serving dish.

评鉴

相传此面点是由一位姓钟的老师傅发明，因色泽金黄、酥脆香糯、外表略带焦香味，故名钟焦粑。

NOTES

This dish is said to have been invented by a man named Zhong. The pies are golden brown, crispy and aromatic.

09 钟焦粑 ZHONG'S CRISPY PIE

10 农夫饼

FARMER'S PANCAKE

食材配方

面粉200克　　玉米粉200克
糯米粉20克　　黄豆粉20克
泡打粉3克　　鸡蛋200克
白糖50克　　　清水600毫升

制作工艺

1. 将面粉、玉米粉、糯米粉、黄豆粉、泡打粉、鸡蛋、白糖、清水入盆调匀，醒发15分钟，制成粉浆。
2. 平底锅置火上，放油烧至120℃，舀入粉浆，烙成饼状，煎至两面金黄后装盘即成。

评鉴

农夫饼是资阳市农村常见的一种风味小吃，以面粉、玉米粉等多种原料制作而成，简单实惠、易于操作。此饼有所改良，制作更精细，色泽金黄，质感软糯适中，味道香甜可口。

INGREDIENTS

200g wheat flour; 200g corn flour; 20g glutinous rice flour; 20g soy bean; 3g baking powder; 200g eggs; 50g sugar; 600ml water

PREPARATION

1. Mix about flour, corn flour, glutinous rice flour, soy bean powder; baking powder, eggs, sugar and water. Leave the paste to ferment for 15 minutes.

2. Heat the cooking oil in the pan to 120°C. Add the paste and bake till the two sides of the pancake are golden brown. Serve the dish.

NOTES

Farmer's Pancake is a local dish in countries of Ziyang City and mainly made from rice wheat flour and corn flour, etc. The cooking is operational and has been developed. The pies taste soft, glutinous, sweet and delicate.

11 养生竹燕窝
NUTRITIONAL BAMBOO FUNGUSES

食材配方

竹燕窝200克　小米100克　玉米粒50克
薏米25克　枸杞2克　姜片5克
葱段20克　食盐3克　水淀粉20克
奶汤1 000毫升

制作工艺

1. 奶汤入不锈钢锅，放入小米、玉米粒、薏米、姜片、葱段煮成小米粥，再加入食盐、水淀粉，收汁浓稠后装碗。
2. 竹燕窝用清水浸泡，入奶汤中煮熟，捞出后放在小米粥上，点缀上枸杞即成。

评鉴

竹燕窝为四川特产，是一种生长在竹上的天然菌类，营养丰富，口感细腻，因形似燕窝而得名。将竹燕窝配以奶汤小米粥，味道咸鲜，口感更顺滑，营养更丰富并易于吸收，为养生补益之佳品。

INGREDIENTS

200g bamboo funguses; 100g millets; 50g corn grains; 25g coixseeds; 2g wolfberries; 5g ginger, sliced; 20g spring onions, segmented; 3g salt; 20g batter; 1,000g milky stock

PREPARATION

1. Combine milky stock, millets, corns, coixseeds, ginger and spring onions in a stainless steel pot, and simmer till all ingredients are cooked through. Add salt and batter to the porridge, simmer till thick, and ladle to bowls.

2. Soak the bamboo funguses in water, boil in milky stock till cooked and transfer to the bowls on the surface of the porridge. Garnish with wolfberries.

NOTES

Bamboo funguses, which have high nutritional values, are also called bamboo bird's nests because of their resemblance in appearance. The dish is a nutritious combination of funguses and milky millet porridge, benefiting your health and stimulating your tasting bud.

12 脆炸蜜荷花
DEEP FRIED LOTUS FLOWERS

食材配方

鲜荷花瓣100克　　脆浆糊100克
干辣椒粉碟1个　　蜂蜜20毫升
食用油1 000毫升（约耗30毫升）

制作工艺

1. 鲜荷花瓣放入淡盐水中洗净、凉干。

2. 锅置火上，放油烧至150℃，将鲜荷花瓣挂上脆浆糊后入锅炸至金黄，捞出装盘，配上干辣椒粉碟、蜂蜜即成。

评鉴

荷花以其"出淤泥而不染"之品格为世人称颂，有美容养颜、清心凉血、清热解毒的作用。此菜以荷花为食材炸制而成，美观大气，口感酥脆，香甜可口。

INGREDIENTS

100g lotus pedals; 100g crispiness paste; 1 ground chili saucer; 20ml honey; 1,000ml cooking oil (about 30ml to be consumed)

PREPARATION

1. Rinse the lotus pedals in salted water, and drain.

2. Heat oil in a wok to 150℃, coat the lotus pedals with crispiness paste and deep fry till golden. Transfer to a serving dish, and serve with the saucer and honey.

NOTES

Lotus is known for its symbolism of integrity and benefits in beauty and health preservation. This crispy lotus dish is pleasant both to the eye and the mouth.

INGREDIENTS

120g edible mulberry leaves; 100g thin Chinese style dry noodles; 5g green peppers, finely chopped; 5g red peppers, finely chopped; 35g cornstarch; 35g self-rising wheat flour; 35g wheat flour; 4g salt; 120ml water; 2g ground Sichuan pepper; 5 rose petals; 1,000ml cooking oil (about 60ml to be consumed)

PREPARATION

1. Mix cornstarch, self-rising flour, wheat flour, salt and water, blend well to make crispiness paste.

2. Mix the Chinese style dry noodles well with crispiness paste, pour in stainless steel strainer to form like a bird nest, then deep-fry in oil until crispy and ladle out.

3. Heat oil in a wok to 130°C, coat the mulberry leaves with crispiness paste, then deep-fry until crispy, ladle out and transfer to the bird nest, sprinkle with Sichuan pepper, add rose petals, green peppers and red peppers then serve.

13 桑叶酥

MULBERRY LEAF FRITTERS

食材配方

食用桑叶120克	细干面条100克
青椒丁5克	红椒丁5克
生粉35克	自发面粉35克
面粉35克	食盐4克
清水120毫升	花椒粉2克
玫瑰花瓣5瓣	食用油1000毫升（约耗60毫升）

制作工艺

1. 将生粉、自发面粉、面粉、食盐加清水调成脆浆糊。
2. 将细干面条与脆浆糊拌匀，入不锈钢漏碗定型成鸟巢，入油锅中炸酥后捞出。
3. 锅置火上，放油烧至130℃，将桑叶裹上脆浆糊，入油锅中炸至酥脆，捞出后装入鸟巢，撒上花椒粉，加上玫瑰花瓣、青椒丁、红椒丁即成。

NOTES

Mulberry leaves are one of the major products in Lezhi County. Local people prefer to cook the mulberry leaves by deep-frying, which has been a local cuisine tradition. This dish has pleasant green color, crispy and appetizing taste, salty, savory and slightly numbing flavor.

评鉴

乐至是产桑大县，将桑叶炸制成菜已经成为当地富有特色的饮食传统。此菜色泽碧绿，质地酥脆爽口，味道咸鲜微麻。

14 桑叶薄脆饼
CRISPY MULBERRY LEAF PANCAKE

INGREDIENTS
150g premier wheat flour; 1 egg; 20ml edible mulberry leaves juice; 30g sweet potato cornstarch; 2g sesames; 4g shallots; 2g salt; 2g sugar; 5ml cooking oil

PREPARATION
1. Add cornstarch, egg, edible mulberry juice, shallots, salt, sugar in flour to blend well to make thick flour paste.

2. Heat a pan over a low flame, brush with the cooking oil, add flour paste to make round pancake, sprinkle with sesames and fry until golden brown, then remove. Fold the pancake in half and cool down then transfer to a serving dish.

NOTES
Mulberry Leaves Crispy Pancake is one of Lezhi County's local snack, which has golden brown color, thin pancake, crispy, sweet and salt balance taste and unique mulberry leaves fragrance.

食材配方

特级面粉150克	鸡蛋1个
食用桑叶汁20毫升	红薯淀粉30克
白芝麻2克	香葱4克
食盐2克	白糖2克
食用油5毫升	

制作工艺

1 特级面粉中放入红薯淀粉、鸡蛋、食用桑叶汁、香葱、食盐、白糖，调匀成浓稠状面糊。

2 平底锅置小火上，抹上食用油，倒入面糊，制成圆形薄饼，撒上白芝麻，煎成金黄色，起锅后折成扇形，冷却后装盘即成。

评鉴

桑叶薄脆饼是乐至县的一道特色小吃，色泽金黄，薄如蝉翼，口感酥脆，味道咸甜适中，有桑叶独特的清香。

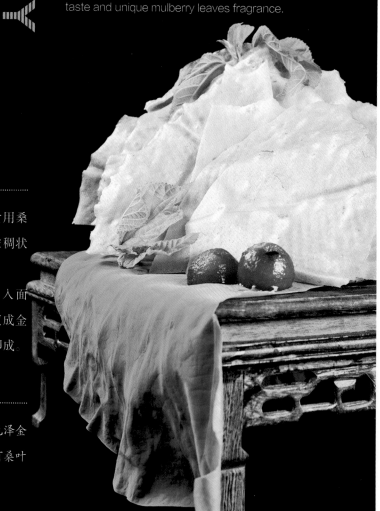

15 天池藕粉

TIANCHI LOTUS ROOT PORRIDGE

食材配方

原味藕粉150克

白糖20克

蜜饯20克

凉开水150毫升

沸水300毫升

制作工艺

1 蜜饯切成细粒。

2 藕粉、白糖加凉开水搅匀，再加入沸水搅拌成糊状，最后加入蜜饯搅匀即成。

评鉴

天池藕粉是国家地理标志产品和四川省名牌产品，色泽白中微红、细腻滑润，用沸水冲调后呈半透明胶糊状。味道甘甜香醇，具有清心明目、美容养颜等作用，尤其适用于老幼人群。

INGREDIENTS

150g plain lotus root powder; 20g sugar; 20g preserved fruits; 150ml cold boiled water; 300ml boiling water.

PREPARATION

1. Finely chop the preserved fruits.

2. Mix well the lotus root powder, sugar and cold boiled water in a bowl. Then add boiling water and stir it until it looks pasty. Add preserved fruits and mix all the ingredients well.

NOTES

Tianchi Lotus Root Powder is both the product of geographical indications and a Sichuan famous brand. It is white and reddish in color and it tastes fine and smooth. After being brewed by the boiling water, it becomes semitransparent paste with mellow and sweet flavor, which benefits heart and eyesight and helps to maintain beauty and keep young. It best fits the old and the young.

16 荷叶莲子粥
MILLET PORRIDGE WITH LOTUS LEAVES AND SEEDS

食材配方

小米250克　　鲜莲子100克
鸡汤1000毫升　食盐6克
新鲜荷叶1张

制作工艺

1. 新鲜荷叶切碎。
2. 锅置火上，入小米、鲜莲子，掺入鸡汤，大火烧沸后，改用小火煮至成粥，放入碎荷叶煮5分钟，加入食盐后装碗即成。

评鉴

荷叶具有消暑健脾、散瘀止血的功效，新鲜莲子清香可口。将新鲜荷叶、新鲜莲子与暖胃的小米熬成粥，咸鲜清香，在炎热的夏季食用更能增添食欲。

INGREDIENTS

250g millet; 100g fresh lotus seeds; 1,000ml chicken stock; 6g salt; 1 fresh lotus leaf

PREPARATION

1. Finely chopped the fresh lotus leaves.

2. Place millet, fresh lotus seeds, and chicken stock in the wok and heat with high heat. After the soup starts to boil, simmer till they become porridge. Place lotus leaves into the porridge and simmer for 5 minutes. Add salt, stir well and serve.

NOTES

Lotus leaves can help tonify spleen, dissipate blood stasis and staunch. The fresh lotus seeds are delicious. With fresh lotus leaves, lotus seeds and millet, this dish is salty and fresh. Having it in summer can give people a good appetite.

Chapter Four

ZIYANG FAMOUS SPECIAL FOOD STREETS (TOWNS) AND RESTAURANTS

Since ancient times, agriculture has been well-developed in Ziyang. As is located in the traffic arteries of Chengdu and Chongqing, there are lots of travelling merchants and frequent commerce and trade. With numerous restaurants along the streets and rich varieties of food, it provides places for local people and traveling people to enjoy delicious food, which promote the development of food industry of Ziyang and make important contribution to inheriting the culture of Sichuan cuisine. Today, under the background of "Internet +", with the rapid development of modern service industry and tourism, and with the constant improvement of people's needs for a better life, Ziyang Municipal Party Committee and Municipal Government attach great importance to and support the development, transformation and upgrading of food industry, making great efforts to build special food brand. On the one hand, the government actively cultivates leading enterprises, develops small and medium-sized enterprises, constantly raises the informationalized level of food industry, and improves the quality of food products and services, thus creating a number of Chinese time-honored brands and lots of catering enterprises and shops well-known around Sichuan and even the whole country. On the other hand, the government actively develops food tourism resources and builds food tourism brands, famous special food streets and towns keeping popping up. These special food streets, small towns, catering enterprises and shops of Ziyang have become ideal places for people to pursue and enjoy delicious food, become important carriers to spread and carry forward the food culture of Ziyang, and become beautiful business cards of Ziyang food culture tourism. Due to space constraints, the authors will only select some of the famous special streets (towns) and restaurants for brief introductions.

第四篇 资阳

名特美食街（镇）与餐饮店

资阳自古以来农耕发达，又位于成渝两地的交通要冲，往来商旅不断，商贸频繁，餐饮店铺沿街林立，美食品种丰富，为当地百姓和往来商旅人提供了品尝美食的场所，推动了资阳美食行业的发展，为传承川菜文化做出了重要贡献。

时至今日，在『互联网＋』的时代背景下，随着现代服务业、旅游业的快速发展和人们对美好生活需求的不断提升，资阳市委、市政府高度重视并支持美食产业的发展和转型升级，努力打造特色美食品牌，一方面积极培育龙头企业，发展壮大中小企业，不断提高美食产业信息化水平，提升美食产品品质与服务，拥有了一批中华老字号和四川乃至全国知名的餐饮企业和店铺；另一方面积极挖掘美食旅游资源，打造美食旅游品牌，名特美食街区和小镇不断涌现。而这些资阳美食的名特街区、小镇，以及餐饮企业、店铺已成为民众追寻美食、享受美食的理想去处，既是传播、弘扬资阳美食文化的重要载体，也是资阳美食文化旅游的靓丽名片。由于篇幅所限，这里仅选取其中的一部分名特街（镇）与餐饮店进行简要介绍。

FAMOUS SPECIAL FOOD STREETS AND TOWNS
名特美食街区与小镇

壹 I

雁江三贤餐饮特色街 | THREE SAGES SPECIAL FOOD STREET, YANJIANG DISTRICT 1

雁江三贤餐饮特色街位于资阳市雁江区三贤广场，因纪念资阳古代三贤，即苌弘、王褒、董钧而得名，被评为省级"餐饮安全示范街"。它是一条特色餐饮较为集中的大型餐饮街区，也是雁江区特色饮食的主要区域，全长700米，品位较高、规模较大、功能较全，有从事餐饮及相关行业的商家200余户。该街区环境优美，风格独特，餐饮氛围浓郁，已成为展示雁江区美食文化形象的一张城市名片。

Yanjiang Three Sages Special Food Stree is located in Three Sages Square of Yanjiang District, Ziyang City. The square is named to commemorate three ancient sages, namely, Chang Hong, Wang Bao, and Dong Jun. The street is awarded as provincial "Food Safety Demonstration Street". It is a large catering block with concentrated characteristic foods, and it is also the main area of special catering in Yanjiang District. With a total length of 700 meters, it is a large-scale street with high grade and complete functions, which has more than 200 businesses engaged in catering and related industries. With beautiful environment, unique style and rich catering atmosphere, this block has become a city card to display the food culture image of Yanjiang District.

2 DAGAO COMMERCIAL PEDESTRIAN STREET, YANJIANG DISTRICT | 雁江达高商业步行街

　　雁江达高商业步行街位于资阳市雁江区达高国际中心地段，2012年已成为"电子商务示范街"。它是集特色餐饮、休闲购物、电子商务于一体的综合性步行街，全长约200米，有从事餐饮、百货和电子商务的商家90余户。沿街环境优美，许多特色餐饮店汇集于此，其中最具代表性的餐饮店有烫起泡火锅、绝之味美蛙鱼头等。经过餐饮人多年的创新沉淀，该街区的餐饮氛围浓郁，是资阳市民饮食、休闲、娱乐的理想去处。

　　Yanjiang Dagao Commercial Pedestrian Street is located in Dagao International Center area of Yanjiang District, Ziyang City, which has become the "E-commerce Demonstration Street" since 2012. It is a comprehensive pedestrian street integrating characteristic catering, leisure and shopping and e-commerce, with a total length of about 200 meters and more than 90 businesses engaged in catering, department stores and e-commerce. The environment along the street is beautiful, and many featured restaurants gathering here, among which Tangqipao Hot Pot, Weizhijue Delicious Frog and Fish Head are the most representative restaurants. After years of innovation and accumulation, this block has a rich catering atmosphere, and is an ideal place for citizens of Ziyang to eat, relax and entertain.

雁江凤岭路美食街 | FENGLING FOOD STREET, YANJIANG DISTRICT 3

　　雁江凤岭路美食街是资阳市重点打造的集美食、休闲、娱乐为一体的商业街，也是小吃、中餐、火锅、夜宵百家争鸣的美食一条街。该街区拥有门市400余间，资阳市有名的餐饮企业，如雅香阁、香泽苑、阁里香等均入驻其中。借助紧邻凤岭公园的地理优势和资阳餐饮名店的品牌效应，人气大增，数家大型酒店和10余家KTV也相继开业。经过近10年的发展，该街区的餐饮已形成品种齐全、特色突出的风格，成为展现资阳美食文化、领略资阳美食特色与风情的好地方。

Yanjiang Fengling Food Street is the mainly constructed commercial street by Ziyang City integrating food, leisure, and entertainment. It is also a competitive food street full of various snacks, Chinese food, hot pot, and night snacks. It has more than 400 stores, in which Ziyang's famous catering enterprises, such as Yaxiang Ge, Xiangze Yuan, Gelixiang, are settled there. By virtue of the geographical advantage of close to Fengling Park and the brand effect of Ziyang's famous restaurants, the popularity of the food street is greatly increased, and several large hotels and more than 10 KTV also open one after another. After nearly 10 years of development, the catering in this block has formed a complete variety with prominent characteristics. It has become a good place to show the food culture of Ziyang and to appreciate its food characteristics and customs.

4 NINGDU NEW TOWN PEDESTRIAN STREET, ANYUE COUNTY 安岳柠都新城步行街

　　安岳柠都新城步行街位于安岳县南山新区核心地段，呈"T"形分布，分为景观步行街与特色餐饮广场两个部分，全长约600米，街宽30米，由2栋3层全框架中空式商业体构成。整个街区聚集大型百货、大型超市、影院、专业卖场，以及中高端餐饮娱乐等多种业态，全力打造集休闲、娱乐、餐饮、购物、文化交流等为一体的一站式商业中心，现已逐步成为安岳新城最繁华、最现代的商业圈之一。其中，特色美食店铺有数十家，包括菲美斯酒店、炖香园酒楼安岳旗舰店、帝方品味茶酒楼、老时代火锅、石爆木桶鱼等，满足了不同口味人群的美食需求。

　　Anyue Ningdu New Town Pedestrian Street is located in the core area of Nanshan New District, Anyue County. It is distributed in T-shape, being divided into two parts of landscape pedestrian street and characteristic catering square. The total length is about 600 meters, and the width is about 30 meters. It is composed of two 3-storey full frame hollow commercial buildings. The whole block is full of large department stores, large supermarkets, cinemas, professional stores, as well as high-end catering and entertainment and other types of businesses, to build a one-stop business center integrating leisure, entertainment, catering, shopping, cultural exchange and so on. Now it has gradually become one of the most prosperous and modern business circles in Anyue New City. Among them, there are dozens of special food shops, including Feimos Hotel, Dunxiang Yuan Restaurant Anyue Flagship Store, Difang Pinwei Tea Restaurant, Old Time Hot Pot, Shibao Wooden Barrel Fish, etc., which can meet the needs of different tastes.

安岳中央时代广场美食街区 | CENTRAL TIMES SQUARE FOOD STREET, ANYUE COUNTY | 5

安岳中央时代广场美食街区位于安岳县新城区，包括安岳柠都大道、广场路、九韶路、龙郡66侧街等街道，以中央时代广场商业综合体为中心，汇聚了300余家餐饮酒店、快餐小吃店、茶座饮品店。其中，中央时代广场的总建筑面积56 568.78平方米，入驻各类商家100余户，其中餐饮商家占40%，包括九格子、川道耗儿鱼、蜀大侠火锅等40余家。以时代广场众多餐饮为集聚中心，周边餐饮也发展迅速，汇聚了以柠檬花精品店、知味轩等为代表的280余户各类美食商家，形成了直径500米、总量300余家的美食圈，成为安岳新城区名符其实的特色餐饮集聚地，极大地提升了安岳城市形象。

Anyue Central Times Square Food Street is located in the new urban district of Anyue County, including Anyue Ningdu Avenue, Square Road, Jiushao Road, Longjun 66th Side Street, etc. Centered on the commercial complex of the Central Times Square, it has gathered more than 300 restaurants and hotels, fast food and snack shops, tea houses and beverage shops. Among them, the total commercial building area of the Central Times Square is 56,568.78 square meters, and more than 100 businesses are settled, among which catering businesses account for 40%, including more than 40 restaurants, such as Jiugezi, Chuandao Corydoras and Shudaxia Hot Pot. Taking the numerous dining of the Times Square as the gathering center, the surrounding catering also develops rapidly. It has brought together more than 280 various types of food businesses represented by Lemon Flower Boutique and Zhiwei Xuan, forming a food circle whose diameter is 500 meters, and the total amount of businesses is over 300. Thus it has become the characteristic food gathering place of Anyue new urban district, which has greatly improved the city image of Anyue.

6 SHUAIXIANG AVENUE FOOD STREET, LEZHI COUNTY 乐至帅乡大道美食街区

乐至帅乡大道美食街区位于乐至县南塔新区，由帅乡大道及附近的文庙沟美食街区等构成。帅乡大道因乐至为陈毅元帅的家乡而得名，2012年荣获四川十大最美街道称号，2018年被评为资阳餐饮名街。帅乡大道全长2 000米，宽50米，沿街拥有许多大型餐饮企业和100余家特色小吃店，包括玲路酒店、鑫鹏大酒店、八颗米土菜馆、川家人等。文庙沟美食街区则以民乐路、文庙街为中心，全长2 000米，聚集了香牌坊、凉山风味、一品肥牛等大中小型餐饮企业、火锅店、烧烤店、特色小吃店共150余家，是民众享受美食的理想去处。

Lezhi Shuaixiang Avenue Food Street is located in the Nanta New District of Lezhi County, mainly consisting of Shuaixiang Avenue and the nearby Wenmiaogou Food Street. Shuaixiang Avenue gets its name because Lezhi County is the hometown of Marshal Chen Yi. In 2012, it won the title of "Ten Most Beautiful Streets in Sichuan"; in 2018, it was rated as "Famous Food Street of Ziyang". Shuaixiang Avenue is 2,000 meters long and 50 meters wide. Along the street, there are lots of large catering enterprises and more than 100 special snack shops, including Linglu Hotel, Xinpeng Hotel, Eight Rice Local Dishes, Chuan Family, etc. Wenmiaogou Food Street, centered on Minle Road and Wenmiao Street, is 2,000 meters long. There are over 150 small and medium-sized catering enterprises, hot pot restaurants, barbecue restaurants and special snack shops, such as Xiangpaifang, Liangshan Flavor, Best Quality Fat Cow, etc., which have become ideal places for people to enjoy delicious food.

雁江中和镇：小龙虾之约
A CRAYFISH DATE, ZHONGHE TOWN, YANJIANG DISTRICT

7

雁江中和镇位于资阳市雁江区东北部，是全国重点镇、四川省百镇建设试点行动重点镇、四川省环境优美示范镇。该镇是《白毛女》作者、作家邵子南的故乡，距雁江城东新区13公里，与天府国际机场的直线距离18公里，遂资眉高速、成资渝高速交会于此。中和镇是农业大镇，总面积121平方公里，辖25个村、1个居委会，人口5.8万，耕地面积7.8万亩；它也是雁江长寿之乡核心区，四季景色佳美，旅游资源丰富，花溪谷、罗汉洞等极具旅游开发价值。

Yanjiang Zhonghe Town is located in the northeast of Yanjiang District, Ziyang City. It is a national key town of China, one of the key towns of Sichuan "Hundred Towns Construction Pilot Action", and a model town with beautiful environment in Sichuan Province. The town is the hometown of writer Shao Zinan who is the author of White Haired Girl. It is 13 kilometers away from the East New District of Yanjiang and 18 kilometers as the crow flies away from Tianfu International Airport. Therefore, Zimei Expressway and Chengziyu Expressway meet here. Zhonghe is a major agricultural town, with a total area of 121 square kilometers, 25 villages and 1 neighborhood committee under its jurisdiction, a population of 58,000, and a cultivated land area of 78,000 mu. Besides, it is also the core area of the longevity town of Yanjiang, with beautiful scenery in four seasons and rich tourism resources as Flower Valley, Rohan Cave and some other spots have great tourism development value.

中和镇美食资源丰富，有"榨菜之乡""果蔬之乡""酿造之乡""小龙虾之乡"的美称。现有中和醋厂、国灿榨菜厂、露乐矿泉水、旺鹭食品公司等7家具有一定规模的食品企业；建有干沟万亩大雅柑示范园、巨善皮球桃基地，辐射带动种植晚熟柑橘、皮球桃等1.8万余亩；以玉林食品、国灿榨菜等企业为依托，种植榨菜3万余亩；以当地特色小龙虾餐饮企业为支撑，新增小龙虾养殖3000余亩，形成了"山上果树、山脚蔬菜、田中小龙虾"的立体主导产业。其中，"资阳中和小龙虾""中和大雅柑"和"明月精品果园晚血橙"等申请注册了产品商标或有机认证。"中和农产品"已成为资阳美食不可或缺的重要元素和物质基础，实现了农业品牌与特色餐饮产业互促共融。此外，中和镇在美食旅游方面也成效显著，成功举办了两届小龙虾美食节，吸引游客食客20余万人次，"龙虾旅游小镇"初具雏形，美食品牌再上新台阶。

Zhonghe is rich in food resources, and is known as "Hometown of Preserved Mustard", "Hometown of Fruits and Vegetables", "Hometown of Brewing", and "Hometown of Crayfish". There are 7 food enterprises with a certain scale, such as Zhonghe Vinegar Factory, Guocan Preserved Mustard Factory, Lule Mineral Water, Wanglu Food Company, and so on. It has built Gangou 10,000 Mu Daya Citrus Demonstration Park and Jushan Piqiu Peach Base, which radiate and promote the planting of more than 18,000 mu of late-maturing citrus and piqiu peach. Relying on Yulin Food, Guocan Preserved Mustard and other enterprises, it has planted more than 30,000 mu of mustard. Supported by local crayfish catering enterprises, more than 3,000 mu of crayfish cultivation has been added, forming a three-dimensional leading industry of "fruit trees on the mountain, vegetables at the foot of the mountain, crayfish in the field". Among them, "Ziyang Zhonghe Crayfish", "Zhonghe Daya Citrus", "Mingyue Quality Orchard Late-maturing Blood Orange" and some other products have applied for registration of product trademarks or organic certification. "Zhonghe Agricultural

Product" has become an indispensable element and material basis of Ziyang cuisine, realizing mutual promotion and integration of agricultural brand and characteristic catering industry. In addition, Zhonghe has also achieved remarkable results in food tourism. It successfully has held Crayfish Food Festival twice, attracting more than 200,000 tourists. The "Crayfish Tourism Town" has taken shape, and the food brand has reached a new level.

8 LOTUS FRAGRANCE SUFFUSING, DANSHAN TOWN, YANJIANG DISTRICT 雁江丹山镇：莲藕飘香

雁江丹山镇位于雁江区东部，始建于唐贞观四年（公元630年），如今是四川省现代农业特色小镇和农旅特色小镇。全镇面积119平方公里，总人口6.8万余人，农业和旅游业融合发展、特色突出。

Yanjiang Danshan Town, located in the east of Yanjiang District, was founded in the fourth year of Zhenguan Period of the Tang Dynasty (630 A. D.). It is now a characteristic town of modern agriculture and rural tourism in Sichuan Province. The town covers an area of 119 square kilometers, with a total population of more than 68,000 people. It is featured by integrative development of agriculture and tourism.

丹山镇内有"四个一万亩"产业和"一带五园一基地"的特色旅游路线，已成功举办了两届主题荷花节。其中，前者是指万亩莲藕、万亩核桃、万亩柑橘和万亩蔬菜；后者是指川中莲藕观光带、顺峰荷塘人家观光园、红村藤椒主题公园、桥沟桃花岛湿地公园、白塔文化公园、李伸沟李子采摘园、黄氏宗祠慈孝文化教育基地。丹山镇依托万亩莲藕与荷花节，研发创制莲藕宴，包括荷塘月色、荷叶粥、藕香排骨等，色味俱佳，成为荷花节期间人们的旅游美食首选。此外，丹山镇还有许多特色美食及土特产食材，如胡凉粉远近闻名，百草灰咸鸭蛋油而不腻。核桃、藤椒、青脆李等，也深受当地民众和旅游者的喜爱。

Danshan has two characteristic tourist routes of "Four Ten Thousand Mu" Industry and "One Belt Five Parks One Base", and has successfully held Theme Lotus Festival twice. Regarding to the two routes, the former refers

to 10,000 mu lotus root, 10,000 mu walnut, 10,000 mu citrus and 10,000 mu vegetable; the latter refers to Central Sichuan Tourist Belt of Lotus Root, Shunfeng Lotus Sightseeing Park, Red Village Pepper Theme Park, Qiaogou Peach Blossom Island Wetland Park, White Tower Cultural Park, Lishengou Plum Picking Park, and Huang's Ancestral Hall Love and Filiality Culture and Education Base. Relying on 10,000 mu lotus root and Lotus Festival, Danshan develops and creates lotus root banquet, including Moonlight over the Lotus Pond, Lotus Leaf Porridge, Ribs Stewed with Lotus Roots, etc., which is pleasant to the eye and taste, and has become the first choice for people to taste during Lotus Festival. In addition, Danshan also has many other special delicious foods and local products, such as the famous Hu's Pea Jelly, the non-greasy salted duck eggs. Besides, walnuts, pepper, green crisp plum, etc. are very popular with local people and tourists.

安岳柠檬小镇：柠檬主题乐园　LEMON THEME PARK, LEMON TOWN, ANYUE COUNTY　9

安岳柠檬小镇位于安岳县文化镇，紧邻成安渝高速入口，被评为AAA级旅游风景区，列入国家现代农业示范园、授予"中国柠檬小镇"称号。该小镇由四川宝森农林科技集团投资建设，现已建成6 000亩核心区域，包括柠檬种植基地、果蔬采摘基地、柠檬园康养中心、花卉餐饮接待大厅、宝森果园、醉美荷花池、柠檬音乐广场等，拥有个体经营户约600户。

Located in the Wenhua Town of Anyue County, Anyue Lemon Town is close to the entry of Chenganyu Expressway, is rated as AAA Tourism Scenic Spot, is included in the National Modern Agriculture Demonstration Park, and is awarded the title of "China Lemon Town". Invested and built by Sichuan Baosen Agriculture and Forestry Technology Group, the town has now built a core area of 6,000 mu, including Lemon Planting Base, Fruit and Vegetable Picking Base, Lemon Garden Health Care Center, Flower Dining Reception Hall, Baosen Orchard, Gorgeous Lotus Pond, Lemon Music Square, etc., with about 600 individual business households.

柠檬小镇致力于打造"生态农业+旅游休闲+健康养老+文化创意"于一体的综合农业休闲观光旅游园区，先后成功举办了3场大型音乐节、第十届安岳柠檬节（主会场）开幕式，以及端午文化节、美食节、花卉果蔬采摘节等活动，2017年，全年累计解决就业人数3万余人次，有力地带动了当地经济发展，促进了农民稳定增收。

Lemon Town is committed to building a comprehensive agricultural leisure tourism park integrating "eco-agriculture + tourism and leisure, health and old-age care + cultural creativity". It has successfully held three big music festivals, the 10th Anyue Lemon Festival (Main Venue) Opening Ceremony and the Dragon Boat Festival, Food festival, flower, fruit and vegetable picking festivals, etc. In 2017, more than 30,000 people were employed, which gave a strong boost to local economy and steadily increased rural incomes.

10 DISCOVERY OF TRADITIONAL FOOD CUSTOMS, LABOR TOWN, LEZHI COUNTY 乐至劳动镇：发现传统美食民俗

乐至劳动镇位于乐至县北部，原为复兴场，1935年改复兴乡，1958年改公社，1959年经陈毅元帅提议更名为"劳动公社"，1984年置劳动乡，1992年建镇。该镇总土地面积60.5平方公里，总人口4.2万，不仅是汉代孝子薛苞的故乡，也是陈毅元帅故里。2014年，劳动镇旧居村被列为第三批中国传统村落名录。2019年，劳动镇被评为四川省文化旅游特色小镇。

Located in the north of Lezhi County, Lezhi Labor Town originally was Fuxiang Field. In 1935, it was changed to Fuxing Township, and changed to Commune in 1958. In 1959, it was renamed as "Labor Commune" by Marshal Chen Yi. It was set up as Labor Township in1984 and as Labor Town in 1992. With a total area of 60.5 square kilometers and a total population of 42,000, the town is not only the hometown of Xue Bao, a dutiful son in the Han Dynasty, but also the hometown of Marshal Chen Yi. In 2014, Jiuju Village in Labor Town was included in the third group of List of Traditional Chinese Villages. In 2019, the town was rated as Sichuan Cultural Tourism Characteristic Town.

劳动镇辖区主产水稻、小麦、玉米，兼产油菜籽等。劳动镇老街始建于明代，以汉代孝子薛苞故里闻名。古镇位于青山脚下，一条小河将其一分为二，东岸老街与西岸新街由1959年建造的劳动桥连在一起，新街往下一两公里处便是陈毅故居。古镇保留有百年黄桷树、石板老街、空中转角楼、城隍庙和乐楼以及古代筒车遗址、古盐井遗址等。此外，古镇上较好地保存着传统饮食民俗。民国时期，茶铺沿街林立，陈毅元帅的父亲就曾在镇上开了七八年的茶铺。传统节日有赶庙会、清明会、土地会、中秋、春节等。古镇至今仍保留着"三茶六礼"婚嫁习俗等，传承着吹糖人、剪窗花、编竹席、酿米酒等传统技艺，是发现和领略传统美食民俗的理想去处之一。

Labor Town mainly produces rice, wheat, corn, and rapeseed. The Ancient Street of Labor Town was built in the Ming Dynasty, well-known as the hometown of Xue Bao, a dutiful son in the Han Dynasty. Located at the foot of Green Mountain, the ancient town is divided into two parts by a small river. The old street on the east bank and the new street on the west bank are connected by the Labor Bridge built in 1959. A kilometer or two down the new street is Chen Yi's former residence. The ancient town retains the century-old ficus virens, the flagstone old street, the overhead turn tower, Town God's Temple and Music Tower, as well as Ancient Cylinder Waterwheel Ruins and Ancient Salt Well Site, etc. Moreover, the ancient town well preserves the traditional eating customs. In the period of the Republic of China, there were many tea shops along the streets; Marshal Chen Yi's father opened a tea shop for seven or eight years in the town. Traditional festivals preserved in the town include Temple Fair, Qingming Festival Fair, Land Fair, Mid-Autumn festival, Spring Festival and so on. The ancient town still retains the marriage customs of "three lots of tea and six presents", inheriting the traditional skills of sugar-figure blowing, paper-cut for window, bamboo mat weaving and rice liquor brewing, etc. It is one of the ideal places to discover and appreciate the folk customs of traditional food.

乐至高寺镇：特色美食肥肠鱼 — DELICIOUS FISH WITH PORK INTESTINES, GAOSI TOWN, LEZHI COUNTY 11

乐至高寺镇位于乐至县西部，是资阳市级特色小镇。该镇距离县城18公里，面积74.6平方公里，其中，耕地面积35 770亩，辖31个行政村、1个居委会，人口3.7万余人，森林覆盖率41%；遂资眉、成安渝高速穿境而过，交通便利，因出产水果、美食而闻名，享有"四季果乡、美味高寺"的美誉。

Lezhi Gaosi Town is located in the west of Lezhi County, which is a municipal characteristic town of Ziyang City. The town is 18 kilometers away from the county seat, with an area of 74.6 square kilometers, among which, the cultivated land area is 35,770 mu. With a population of 37,000 and a forest coverage rate of 41%, 31 administrative villages and 1 neighborhood committee are under its jurisdiction. With Zimei Expressway and Chenganyu Expressway passing through and due to the convenient transportation, it is famous for its fruits and delicacies, and enjoys the reputation of "four-season fruit town, delicious Gaosi".

高寺镇不仅出产优质葡萄、桑葚，制作的肥肠鱼也很有名，有"到高寺必吃肥肠鱼"之说。该镇有10余家肥肠鱼店，鱼片鲜嫩，肥肠味美，麻辣浓香，赢得了当地民众和往来食客的青睐。以美食、水果为依托，高寺镇大力发展乡村旅游，连续多年举办桑葚采摘节、葡萄采摘节、小龙虾音乐啤酒节、李花节、钓鱼比赛等活动，吸引了大量游客，带动了当地经济发展，已形成新型现代特色农业。

Gaosi Town not only produces high quality grapes and mulberries, but also is famous for its Fish with Pork Intestines. Here's a saying that "one must eat Fish with Pork Intestines in Gaosi". The town has more than 10 Pig's Intestines Fish restaurants; the fish fillets are fresh and tender, and the pig's intestines are delicious and spicy, earning the favor of local people and travelers. Relying on delicious food and fruits, Gaosi vigorously develops rural tourism. For many years, it has held activities such as Mulberry Picking Festival, Grape Picking Festival, Crayfish Music Beer Festival, Plum Blossom Festival and Fishing Contest, attracting a large number of tourists, driving the local economic development and forming new modern characteristic agriculture.

迎宾大桥
Yingbin Bridge

FAMOUS SPECIAL FOOD ENTERPRISES AND RESTAURANTS
名特餐饮企业与店铺

金迪大酒店 | JINDI HOTEL　1

金迪大酒店位于资阳市雁江区体育路32号，是旅游涉外三星级酒店，于1998年开业，2011年按照四星级标准打造"三贤文化"主题酒店。

Located at No. 32, Tiyu Road, Yanjiang District, Ziyang City, Jindi Hotel is a three-star hotel concerning foreign affairs. It was opened in 1998 and was built as a "Three Sages Culture" theme hotel in accordance with the four-star standard in 2011.

该酒店以着力打造菜香、酒香、茶香、书香、墨香、花香的六香酒店为目标，并结合资阳特有的"三贤文化"，开发了苌弘文化宴、苌弘御酒、王褒圣茶、苌弘御饼、苌弘年货、苌弘礼粽等特色旅游商品。其中，苌弘文化宴荣获"四川文化名宴"称号，苌弘鲶鱼烹调技艺列入资阳市非物质文化遗产保护名录，2012年11月在中央电视台播出。此外，该酒店还在2011年度四川省首届地方旅游特色菜大赛上获得团体金奖、最佳营养配膳奖、最佳创意奖等"六金一银"的优异成绩。

The hotel aims to build a six-fragrance hotel of food, liquor, tea, book, ink and flower. Combining with the unique "Three Sages Culture" of Ziyang, it develops special tourism products such as Changhong Cultural Feast, Changhong Royal Liquor, Wang Bao Sacred Tea, Changhong Royal Cake, Changhong Spring Festival Goods, Changhong Glutinous Rice Dumpling, and so on. Among them, Changhong Cultural Feast was awarded the title of "Famous Sichuan Cultural Feast", and Changhong Catfish cooking technique was included in Ziyang intangible cultural heritage protection list; which were broadcast on CCTV (China Central Television) in November, 2012. In addition, at Sichuan First Local Tourism Specialty Competition in 2011, the hotel got six gold medals and one silver medal such as Group Gold Medal, Best Nutritional Supplement Award, and Best Creative Award.

为了保护三贤文化专利，金迪酒店成功注册了"苌弘鲶鱼""三贤""王褒圣茶""苌弘御酒"4个极具资阳地方特色的著名商标。同时，还荣获四川省优秀旅游星级酒店、资阳首批服务标准化示范单位、雁江区最佳文明单位、旅游诚信经营单位等称号，连续几年被评为全省和全市旅游工作先进单位，树立了良好的企业形象。

In order to protect the patent of Three Sages Culture, Jindi Hotel has successfully registered 4 famous trademarks of great local characteristics, namely, "Changhong Catfish", "Three Sages", "Wang Bao Sacred Tea", "Changhong Royal Liquor". At the same time, it has also won the titles of Excellent Tourist Star Hotel in Sichuan Province, the First Batch of Service Standardization Demonstration Unit in Ziyang, the Best Civilized Unit of Yanjiang District, and the Tourism Integrity Management Unit; and has been rated as the Provincial and Municipal Advanced Unit in Tourism Work for several years in a row, setting up a good corporate image.

2 GREEN BOYA HOTEL | 格林博雅饭店

格林博雅饭店位于资阳市雁江区外环路311号，2003年被资阳市政府授牌为资阳迎宾馆，于2005年全面对外营业，2006年被评定为资阳市餐饮名店，2010年获得资阳市百佳企业殊荣。

Located at No. 311, Outer Ring Road, Yanjiang District, Ziyang City, Green Boya Hotel was awarded as Ziyang Guest Hotel by Ziyang Municipal Government in 2003 and opened to the public in 2005. It was rated as Ziyang Famous Restaurant in 2006, and won the title of Top 100 Enterprises in 2010.

该饭店占地面积230余亩，建筑面积29 000余平方米，是一家集住宿、餐饮、茶楼、娱乐休闲、会议于一体，并以生态园林为主题的饭店，园区绿化覆盖率达90%，有客房、餐厅、会议休闲中心等三栋独立主体建筑。饭店开发了许多深受食客好评的菜点，其中，口袋玉米饼、金牌吊烧鸡、满园春色、萝卜酥、香汤羊柳等菜品被评为资阳市名菜，有的菜品还获得了烹饪大赛金奖。

The hotel covers an area of more than 230 mu, with a construction area of more than 29,000 square meters. It is a hotel with the theme of ecological garden integrating accommodation, catering, tea house, entertainment and leisure,

and conference. The green coverage rate of the garden reaches 90%, with three independent main buildings of guest room, restaurant, conference and leisure center. The hotel has developed many dishes that are well received by diners. Among them, Pocket Tortilla, Gold Medal Hanging Chicken, Garden in Spring, Radish Crisp, Fragrant Soup of Lamb and other dishes were rated as Ziyang famous dishes, and some dishes even won the gold medal in cooking competition.

晨风大酒店 | CHENFENG HOTEL 3

晨风大酒店位于资阳市雁江区晨风路6号，始建于1974年，2015年，按照国家旅游饭店四星级标准装修改造升级。建筑面积8 000余平方米，地处风景秀丽的九曲河畔，是一家集住宿、餐饮、品茗、会议等功能于一体的高品质、精品酒店。

Located at No. 6 Chenfeng Road, Yanjiang District, Ziyang City, Chenfeng Hotel was built in 1974 and was renovated and upgraded according to the four-star standard of national tourist hotel in 2015. The building area of the hotel is over 8,000 square meters. It is situated by the beautiful Jiuqu River, and is a high quality and boutique hotel collection of accommodation, catering, tea, meeting and other functions of high quality, boutique hotel integrating accommodation, catering, tea tasting, conference and other functions.

该酒店拥有客房98间，餐厅宴会大厅高5米，可同时容纳400人就餐，能满足举办不同档次的各类大型宴会需求，9个装修风格独特的雅间可为顾客提供精品川菜和地方特色美食，以苌弘鱼趣、雁江春暖为代表的菜品，不仅造型美观、寓意丰富，而且味美鲜香、老少皆宜。

The hotel has 98 guest rooms. The banquet hall is 5 meters high. It can accommodate 400 people dining at the same time, and can satisfy the demand for all kinds of large banquets of different grades. Besides, there are nine uniquely decorated private rooms providing customers with high-quality Sichuan cuisine and local cuisine. The dishes represented by Changhong Pleasure of Fish and Yanjiang River's Warm Spring are not only beautiful and meaningful, but also tasty and suitable for people of all ages.

绿色雁江，生态资阳
Green Yanjiang District, Ecological Ziyang City

4 JINJIANG SHUHENG HOTEL | 锦江蜀亨大酒店

锦江蜀亨大酒店位于资阳市雁江区娇子大道，于2003年创建，2004年评为四星级旅游饭店，2009年和2011年被四川省卫生厅和四川省食药局授予餐饮服务业A级单位和餐饮服务食品安全示范单位称号。

Located in Jiaozi Avenue, Yanjiang District, Ziyang City, Jinjiang Shuheng Hotel was founded in 2003 and was rated as four-star tourist hotel in 2004. In 2009 and 2011, the hotel was awarded the titles of Grade-A Catering Service Unit and Catering Service and Food Safety Demonstration Unit by Sichuan Health Department and Sichuan Food and Drug Administration.

该酒店的建筑面积2.2万平方米，主楼7层，副楼3层，共有各类客房137间，有多功能宴会大厅、西餐厅、包间等，餐位数840个，还有会议室等，服务设施齐备。该酒店立足于大众口味，发展地方特色餐饮，各种菜点能满足不同层次的消费需求。其中，钟焦粑、临江寺豆瓣熠黄腊丁被评为资阳名菜。

The Hotel has a building area of 22,000 square meters, 7 floors of the main building, and 3 floors of the sub-building. There are 137 guest rooms of all kinds, as well as multi-functional banquet hall, western restaurant, private rooms, 840 seats, and conference rooms. Based on the taste of the public, the hotel develops local characteristic dishes, and has a variety of dishes to meet different levels of consumer demand. Among them, Zhong's Crispy Pie and Braised Yellow Catfish with Linjiangsi chili bean paste are rated as famous dishes in Ziyang.

5 YANGFENG HOME RESTAURANT | 阳丰家园酒店

阳丰家园酒店位于雁江区东岳山下，于2012年开业，2018年被评为四川餐饮名店。

Located at the foot of Dongyue Mountain, Yanjiang District, Yangfeng Home Restaurant opened in 2012 and was rated as Sichuan Famous Restaurant in 2018.

该酒店是一家以婚寿宴为主题的餐饮企业，营业面积6 000余平方米，有5个宴会厅、11个包间，可同时容纳1 800人就餐。酒店不仅在菜品研究上注重传统与创新的结合，推出的风吹手撕鸡、阳丰土鲢鱼等菜品都得到食客好评，也非常注重厨师团队技能的比赛与交流，2019年初还派出厨师参加了在澳大利亚举办的"2019欢乐春节——行走的年夜饭"活动。此外，酒店还组织团队多次参加省、市相关比赛，曾荣获"地方旅游特色团队银奖"，沱江土鲢鱼荣获四川省第二届"地方旅游特色·热菜金奖"，家园福喜临门、芝士焗红薯等被评为资阳名菜、资阳名小吃。

The restaurant is a catering enterprise mainly holding wedding and birthday banquets, which has a business area of more than 6,000 square meters, 5 banquet halls and 11 private rooms. It can accommodate 1,800 people dining at the same time. The hotel not only pays attention to the combination of tradition and innovation in food research, Wind Dried Hand-Shredded Chicken, Yangfeng Local Silver Carp and other dishes have been well received by diners, but also focuses on the skills competition and communication of chef team. In early 2019, chefs were sent to participate in the "2019 Happy Spring Festival-Walking New Year's Eve Dinner" activity held in Australia. In addition, the hotel has organized teams to participate in provincial and municipal relevant competitions for many times, and won the "Silver Prize of Local Tourism Characteristic Team"; Tuojiang Local Silver Carp won the second Sichuan "Local Tourism Specialty · Gold Medal for Hot Dish"; Home Happiness, Baked Sweet Potato with Cheese and some other dishes were rated as Ziyang Famous Dishes and Ziyang Famous Snacks.

帝景酒楼 | DIJING RESTAURANT 6

帝景酒楼位于资阳市雁江区娇子大道463号，于2006年开业，2011年被评为资阳市服务行业百家优秀企业及资阳市诚信纳税企业，2013年和2015年被评为资阳市名店。

Located at No. 463, Jiaozi Avenue, Yanjiang District, Ziyang City, Dijing Restaurant opened in 2006. In 2011, it was rated as One Hundred Outstanding Enterprises in Ziyang Service Industry and Ziyang Honest Taxpaying Enterprise. In 2013 and 2015, it was awarded as Ziyang Famous Restaurant.

该酒楼是一家以海鲜火锅为品牌、以精品川菜为特色的餐饮企业，拥有可承接婚寿宴的大型宴会厅、多功能会议接待厅和12个环境优雅的包间，可容纳近1000人同时就餐。它坚持"以客为尊，经营多元化，菜品营养养生"的经营理念，引入港式火锅，集众家之长，让每位顾客都能品尝到营养丰富的美味佳肴，其特色菜品有一品酱香鲍、石盘牛仔骨、粉丝捞鹅掌等。

The restaurant is a catering enterprise with seafood hotpot as its brand and delicate Sichuan cuisine as its characteristic, which has large banquet hall for wedding and birthday banquets, multi-functional conference reception hall and 12 environment elegant private rooms. It can accommodate 1,000 people dining at the same time. For years, it adheres to the business philosophy of "customer-oriented, diversified operation, and nutritional and healthy dishes". It introduces Hong Kong-style hotpot and brings together the advantages of various restaurants, so that every customer can taste nutritious and delicious food. Its special dishes include Best Quality Sauced Abalone, Stone Plate Beef Short Rib, Goose Web with Silk Noodles, etc.

茂林大酒店 | MAOLIN HOTEL 7

茂林大酒店位于资阳市雁江区茂林大厦，其前身是菜根园酒店，开业于2003年。茂林大酒店被评为资阳市餐饮名店，荣获川菜辉煌30年卓越企业奖。

Established in 2003, Maolin Hotel is located in Maolin Building, Yanjiang District, Ziyang City, which was formerly known as Caigenyaun Hotel. Maolin hotel was rated as Ziyang Famous Restaurant and won Sichuan Cuisine Brilliant 30 Years Outstanding Enterprise Award.

该酒店是一家集餐饮、茶楼、商务于一体的餐饮企业，建筑面积3 800平方米，包括美食餐厅、宴会厅及包间，可同时容纳1 200人就餐，在接待大中型婚寿宴上独树一帜。酒店拥有技术力量较强的厨师团队，以传统川菜为基础，不断改良创新菜点，其特色菜品有沱江河鲤、手撕黄牛肉、仔姜土鳝段等菜品，深受食客喜爱。

The hotel is a catering enterprise integrating catering, teahouse and business, with a construction area of 3,800 square meters, including gourmet restaurant, banquet hall and private rooms, which can accommodate 1,200 people dining at the same time and has a unique style in the reception of large and medium-sized wedding and birthday banquets. The hotel has a chef team of strong technical strength. Based on traditional Sichuan cuisine, it constantly improves and creates new dishes. The special dishes include Tuojiang River Carp, Hand-shredded Yellow Beef, Local Eel Period with Tender Ginger and so on, which are deeply loved by the diners.

8　FANGYUAN HOTEL　方圆宾馆

方圆宾馆位于资阳市雁江区学苑路463号，于2014年开业，曾荣获川菜辉煌30年卓越企业奖和资阳市名店等荣誉称号。

Fangyuan Hotel is located at No. 463 Xueyuan Road, Yanjiang District, Ziyang City. It opened in 2014 and once won Sichuan Cuisine Brilliant 30 Years Outstanding Enterprise Award and Ziyang Famous Restaurant and other honorary titles.

该宾馆是以红色文化和长寿文化打造的主题文化酒店，有餐厅、茶楼、客房、会议室、培训中心等。宾馆不断挖掘地方特色文化，创制了许多菜点品种，在四川省第二届地方旅游特色菜大赛中荣获7个金奖、2个银奖，还荣获了四川省第三届川菜创新大赛团体银奖、四川省第五届烹饪职业技能大赛团体金奖、资阳市第二届民间特色厨艺大赛团体一等奖。其特色菜肴有养生竹燕盅、长生仙果等，前者评为四川省名菜。

It is a theme culture hotel of red culture and longevity culture, including restaurant, teahouse, rooms and conference rooms, training center, etc. It constantly excavates local characteristic culture and creates a lot of dishes, which won 7 gold medals and 2 silver medals in Sichuan 2nd Local Tourism Specialty Competition, and also won a Group Silver Award in Sichuan 3rd Sichuan Cuisine Innovation Contest, Group Gold Award in Sichuan 5th Culinary Vocational Skills Competition, as well as Group First Prize in Ziyang 2nd Folk Characteristic Cooking Competition. Its specialties include Health Preservation Bamboo Swallow Cup, Longevity Fruits and so on, the former which is rated as Sichuan Famous Dish.

9　XIANGZEYUAN RESTAURANT　香泽苑酒楼

香泽苑酒楼位于雁江区凤岭路世纪城宝莲广场附近，于2009年开业，是一家以精品川菜、家常菜、粤菜为主要特色的综合性酒楼。

Xiangzeyuan Restaurant, located near Baolin Square of Century City, Fengling Road, Yanjiang District, was opened in 2009. It is a comprehensive restaurant with delicate Sichuan cuisine, home cooking and Cantonese cuisine as its main characteristics.

该酒楼的建筑面积约5 000平方米，豪华包间20多个，宴会大厅可同时容纳1 000人就餐，并设有空中茶坊。酒楼坚持原料原产地采购和标准服务，专注于制作精品川菜和本地菜，菜品繁多，尤其擅长河鲜、野菜类菜肴的制作，先后推出的香泽醉虾、风情花椒鸡、双豆烧甲鱼等特色菜品深受食客喜欢。

Its gross area is approximately 5,000 square meters, with more than 20 luxury private rooms. The banquet hall can accommodate 1,000 people dining at the same time. And it has an air tea house. The restaurant insists on purchasing raw materials from the origin and on the standard service. It specializes in making various dishes of fine Sichuan cuisine and local cuisine, especially good cooking river fresh and wild vegetable dishes. The new dishes such as Tasty Drunk Shrimp, Flavored Sichuan Pepper Chicken and Braised Turtle with Two Kinds of Beans are very popular.

阳光谷屋时尚餐厅 | SUNSHINE VALLEY HOUSE FASHION RESTAURANT 10

阳光谷屋时尚餐厅位于资阳市雁江区九曲河示范段，创立于1998年，从一个小店经过20余年的匠心锤炼，现今已成长为资阳餐饮界的知名企业。

Located in the demonstration section of Jiuqu River, Yanjiang District, Ziyang City, Sunshine Valley House Fashion Restaurant was founded in 1998. After over 20 years of accumulation and improvement, it has grown into a well-known catering enterprise in Ziyang from a tiny shop.

该餐厅拥有阳光谷屋总店、阳光谷屋小菜馆等4家店，总经营面积达4 000余平方米，可同时容纳1 200人就餐。阳光谷屋以经营时尚川菜为主，厨师团队用最传统的原料烹制出家常菜、网红菜、特色菜等，深受食客喜爱。其代表菜品之一的"炟爪爪"一直是逢桌必点的菜肴；农夫饼软糯香甜，农家风情突出；白色恋人、海鲜粥等时尚新菜受到食客追捧。

The restaurant consists of four stores, including Sunshine Valley House General Store and Sunshine Valley House Small Restaurant. It has a total operating area of more than 4,000 square meters, which can accommodate 1,200 people dining at the same time. Sunshine Valley House is mainly engaged in fashionable Sichuan cuisine. Using the most traditional raw materials, the chef team cooks home-made dishes, web popular series and special dishes which are loved by diners. One of its representative dishes "Very Soft Chicken Claw" is always a must at every table; Farmer's Pancake soft and sweet, with prominent rural flavor; White Lover, Seafood Porridge, and other fashionable new dishes are sought after by diners.

明苑湖休闲农庄 | MINGYUAN LAKE LEISURE FARM 11

明苑湖休闲农庄位于资阳市雁江区保和镇晏家坝，占地面积1446.5亩，于2010年营业，年接待游客13.5万人次，曾获得四川省五星级农家乐、国家级休闲农业与乡村旅游示范点、资阳市十佳农业产业化龙头企业、资阳市乡村旅游十大乡村名店、资阳餐饮名店等荣誉称号。

Opened in 2010, Mingyuan Lake Leisure Farm is located in Yan's Ba, Baohe Town, Yanjiang District, Ziyang City, covering an area of 1,446.5 mu. It receives 135,000 visitors a year, and has been awarded the titles of Sichuan Five-star Farm, National Leisure Agriculture and Rural Tourism Demonstration Site, Ziyang Top 10 Agricultural Industrial Leading Enterprise, Ziyang Top 10 Famous Rural Shops for Rural Tourism, Ziyang Famous Restaurant, etc.

该农庄集旅游、观光、休闲、度假、特种养殖、生态种植和新型农民培训于一体，现有景点5处，包括风情果园、开心农场、生态养殖园等。餐饮中心可同时容纳1 600人进餐，主要烹制农家特色菜，代表性品种有明苑拌土鸡、家常鲫鱼、农家香碗、酸汤乌鱼片、野菜饼等。

The farm integrates tourism, sightseeing, leisure, vacation, special breeding, ecological planting and modern peasant training. There are 5 scenic spots, including Amorous Orchard, Happy Farm and Ecological Breeding Park, etc. Catering center can accommodate more than 1,600 people dining at the same time, mainly cooking farm specialties; the representative dishes are Mingyuan Local Chicken Salad, Home-made Crucian, Farm Incense Bowl Dish, Cuttlefish Slices in Sour Soup, Wild Vegetable Pie, etc.

12 LEMON FLOWER BOUTIQUE HOTEL | 柠檬花精品酒店

柠檬花精品酒店位于安岳县柠都新城中心地段，创建于2013年，是一家将安岳地域文化和现代都市休闲风格有机融合的城市精品酒店。

Located in the center of Anyue Ningdu New City, and founded in 2013, Lemon Flower Boutique Hotel is an urban boutique hotel that organically integrates the regional culture of Anyue and modern urban leisure style.

该酒店设有客房和餐厅、茶楼等，功能及设施齐全，装饰典雅，别具一格。餐厅能同时容纳近千人就餐，茶楼能同时容纳200人品茗。酒店本着"老菜新作，传统菜精做"的原则和饮食与健康相结合的产品加工思路，注重挖掘和采用资阳特色食材和鲜品食材，不断开发适合顾客需求的菜品，形成了柠

檬系列菜品、老家菜系列菜品和时尚创意菜品三大系列。在菜品制作上注重细节、精心制作并且实施标准化，力求质量稳定，在盘饰上突出个性。其较为适销和具代表性的菜品有柠檬酸酸鸡、干豇豆烧野猪肉、柠香牛排、坛香滑肉等。

The hotel is equipped with guest rooms, restaurant, tea house, etc., with complete functions and facilities, elegant decoration, and unique style. The restaurant can accommodate nearly one thousand people dining at the same time, while the teahouse can accommodate 200 people tasting tea at the same time. In line with the principle of "old dish newly cooking, traditional dish finely cooking" and the product processing idea of combining diet with health, the hotel pays attention to the finding and using Ziyang local food materials and fresh food materials, forming three major series of Lemon Series Dishes, Hometown Series Dishes, and Fashionable and Creative Dishes. The dishes focus on details and are elaborately made, and the standardization is implemented, striving for stable quality and distinct decoration. There are over 10 marketable and representative dishes include Lemon Sour Chicken, Braised Free Range Pork with Dried Long Beans, Lemon-Flavored Steak, Altar Slippery Meat, and so on.

夜色中的安岳县中央时代广场
Anyue Central Times Square at Night

13　ANYUE HOTEL　安岳宾馆

安岳宾馆位于安岳县正北街繁华的商业地段，始创于1952年。2005年被评为三星级宾馆，曾荣获资阳市诚信旅游示范单位、资阳市服务业百佳企业和四川省服务标准化单位等称号。

Anyue Hotel is located in the prosperous business district of Zhengbei Street, Anyue County. It was founded in 1952 and was rated as three-star hotel in 2005. Moreover, it has won the titles of Ziyang Integrity Tourism Demonstration Unit, Ziyang Top 100 Service Enterprises and Sichuan Service Standardization Unit.

该宾馆是一家集住宿、餐饮、商务、旅游、会议、娱乐、休闲为一体的综合性三星级宾馆，建筑新颖，装饰典雅，由北楼、东楼组成，拥有客房共154间，餐饮部宴会厅可接待550人同时进餐，14个各具特色的高档雅间可供客人选择，拥有大、中、小型多功能会议室3个，配备了旅游商务中心、茶坊、大堂吧等多种服务设施，能够满足各类型会议及宴会需要。该宾馆的餐饮技术力量较强，制作出了许多资阳地方特色浓郁的菜肴。

It is a comprehensive three-star hotel with new buildings and elegant decoration integrating accommodation, catering, business, tourism, conference, entertainment, and leisure. Consisting of North Building and East Building, it has 154 guest rooms, one banquet hall that can host 550 people dining at the same time, 14 distinctive high-end private rooms for guests to choose, 3 large, medium and small multi-function meeting rooms, as well as various service facilities such as tourism business center, teahouse, lobby bar to meet the needs of all types of meetings and banquets. The hotel has strong catering skills, and has produced a lot of local dishes with rich characteristics of Ziyang.

14　GOURMET RESTAURANT　美食美味大酒店

美食美味大酒店位于安岳县城区，于1996年创建，最初位于资阳县卫生局内，1999年迁至安岳县正北街98号（原新安餐厅旧址），曾荣获四川省餐饮名店、资阳市餐饮名店、资阳市百佳企业等称号。

Located in the urban area of Anyue County, Gourmet Restaurant was founded in 1996. It was originally located in Health Bureau of Ziyang. In 1999, it moved to No. 98 Zhengbei Street, Anyue County (the former site of Xin'an Restaurant). It has won the titles of Sichuan Famous Restaurant, Ziyang Famous Restaurant and Ziyang Top 100 Enterprises.

该酒店多年来为社会解决下岗和待业、失业人员近千人次，培养川菜厨师近500人次，曾荣获中国川菜金奖、技术能手造就奖。2009年，该酒店组织技术力量以柠檬为主研发了系列柠檬菜品，并组合成柠檬风味宴，被评为中国四川名宴。除此之外，美食美味大酒店还制作和经营许多特色菜品，如野菌炖墨鱼、乡村酸鲊肉、阿婆鲊肥肠、米卷回锅鱼、灯影牛肉等，均受到消费者的喜爱。

Over the years, the restaurant has offered jobs to nearly 1,000 persons of laid-off and unemployment for the society, and has trained nearly 500 Sichuan cuisine chefs, obtaining China Sichuan Cuisine Gold Medal and Technical Experts Building Award. In 2009, the restaurant organized technical force to develop a series of lemon dishes, and combined into a Lemon Flavor Banquet which was rated as Sichuan Famous Banquet in China. In addition, the Gourmet Restaurant also produces and operates many specialty dishes, such as Cuttlefish Stewed with Wild Mushroom, Countryside Suanzha Pork, Grandma Fried Pig Intestines, Twice Cooked Fish with Rice Roll, Dengying Beef Slices, etc., which are loved by consumers.

恒和大酒店 | HENGHE HOTEL　　15

恒和大酒店位于安岳县东大街，开业于2014年，是一家集餐饮、住宿、品茗、棋牌娱乐等于一体的高品质豪华型酒店，按国家四星级标准修建和管理，被评为资阳名店。

Located in the Dongda Street of Anyue County, Henghe Hotel was opened in 2014. It is a luxury hotel with high quality, integrating catering, accommodation, tea tasting and chess and card entertainment. It is built and managed according to the national four-star standard, and is rated as Ziyang Famous Hotel.

该酒店建筑面积约16 000平方米，有客房102间，餐饮部约3 000余平方米，拥有大小包间7个，大厅能同时容纳70桌宴席，装修豪华典雅，服务贴心，代表菜品有柠檬手撕兔、板栗爆乳鸽、宫保龙虾等众多菜品，深受消费者欢迎。

The hotel covers an area of 16,000 square meters, with 102 guest rooms and a catering department over 3,000 square meters. There are 7 private rooms of different sizes. The hall can accommodate 70 tables at the same time. It has luxurious and elegant decoration, and considerate service. The representative dishes are Lemon-Flavored Hand Shredded Roast Rabbit, Quick-fry Squab with Chinese Chestnut, Kung Pao Lobster, etc., which are very popular with consumers.

石秀吞大酒店 | SHIXIUTUN RESTAURANT　　16

石秀吞大酒店位于安岳县奎星街22号，创建于2014年，以店名奇、菜品优、环境雅而成为当地著名的餐饮品牌。"石"谐音食，代表安岳石刻历史悠久，"秀"代表安岳风景优美秀丽，"吞"指菜品秀色可餐。

Located at No. 22 Kuixing Street, Anyue County, Shixiutun Restaurant was founded in 2014. It is a famous local catering brand with unique name, excellent dishes and elegant environment. "Shi" means "stone" and sounds like "food" in Chinese, on behalf of the long history of Anyue stone carvings; "Xiu" means "excellent", representing Anyue's beautiful scenery; "Tun" means "swallow", referring to the delicious and tasty dishes.

该酒店经营面积2 000余平方米，精致高雅，舒适温馨，大厅可举办千人宴席，最大包间有26个餐位。酒店经营的菜品特色较为突出，其代表性菜品有牡丹鱼片、麻辣诱惑蛙、清汤鱼丸等，深受食客喜爱。

The restaurant covers an area of more than 2,000 square meters, which is exquisite, elegant, comfortable and warm. The hall can hold a thousand people banquet, and the largest private room has 26 seats. The food features here are prominent; the representative dishes are Peony Fish Slice, Spicy Tasty Frog, Consomme Fish Balls, etc., which are very popular with diners.

周礼伤心凉粉店 | ZHOULI SPICY PEA JELLY SHOP　　17

周礼伤心凉粉店位于安岳县周礼镇，是一家以制作、销售凉粉为特色的老字号小吃店，被评为四川名小吃、资阳市名小吃、资阳市特色餐饮、安岳名菜点，也是"中华老字号"提名品牌，先后被安岳电视台、资阳电视台、四川电视台和中央电视台等宣传报道。

Located in Zhouli Town of Anyue County, Zhouli Spicy Pea Jelly Shop is a time-honored snack store featured by making and selling pea jelly. It has been named as Sichuan Famous Snack, Ziyang Famous Snack, Ziyang Special Catering, and Anyue Famous Dish; it is also the nominated brand of "China Time-honored Brand". Besides, it has been publicized and reported by Anyue TV, Ziyang TV, Sichuan TV and CCTV, etc.

该店的周礼伤心凉粉创制于20世纪初，由文江源在保持传统黄凉粉风味基础上进行改进，研制出独特的调料配方，使凉粉更加香辣可口。20世纪80年代，其凉粉制作技艺被传于其女婿姚长明，得以研发创新，品质不断精进，滋味更胜从前，从而创立了"姚记周礼伤心凉粉"品牌。该凉粉的特点是麻、辣、香、脆，其主料选用上等豌豆，其配料芝麻、花生、香油等由本地生产，其凉粉品种有白油、微辣、中辣、特辣等，不仅是闲暇时人们所钟爱的小吃，还作为一种家常菜被人们端上餐桌，并逐渐成为人们出行、走亲访友、宴席待客之常见美食。

Created in the early 20th century, Zhouli Spicy Pea Jelly in the shop was improved by Wen Jiangyuan on the basis of maintaining the flavor of traditional yellow pea jelly, and a unique seasoning recipe was developed to make it spicier and more delicious. In the 1980s, the pea jelly making skills were passed on to his son-in-law Yao Changming, who achieved some development and innovation, continuously improved the quality, made it taste better than before, thus founded the brand of "Yao's Zhouli Spicy Pea Jelly". Selecting high quality peas as main material and local sesame, peanut, sesame oil as ingredients, the pea jelly has characteristics of being spicy, hot, appetizing, and crisp. The pea jelly has various flavors, such as not spicy, mild spicy, medium spicy, very spicy, etc. It is not only the snack people love to eat in spare time, but also is served as a home-made dish on the table. And gradually, it has become a common dish during travel, visiting relatives and friends, and treating guests at banquets.

18 ANYUE WUMING BRAISED CHICKEN SHOP 安岳吴名烧鸡店

安岳吴名烧鸡店的老店暨总店位于安岳县文化镇新民桥，创立于1988年，另外在岳阳镇还开设有多家分店，以一款无名烧鸡闻名远近。2018年，该店在四川省质量监督品质推广活动中被评为特色餐饮名店。

Located in Xinmin Bridge, Wenhua Town, Anyue County, the old store and general store of Anyue Wuming Braised Chicken Shop was founded in 1988. Opening many branches in Anyue, Anyue Wuming Braised Chicken Shop is renowned because of the Wuming Braised Chicken. In 2018, the store was rated as Famous Specialty Store in Sichuan Quality Supervision and Promotion Activity.

该店的无名烧鸡特色鲜明，将新疆风味与四川特色相结合，由农家土鸡、辣味小辣椒、清香浓郁的大辣椒为食材，加上自制的特色豆瓣酱和川人最喜爱吃的泡辣椒、泡姜，通过炒、烧、炖等烹制而成，被评为资阳名菜。同时，该店的"一鸡三吃"，即泡椒炒鸡杂、鸡杂粉条和鸡血酸汤，味道鲜美，回味无穷，均受到消费者青睐。

The Wuming Braised Chicken has distinctive characteristics, which combines Xinjiang flavor with Sichuan features. It is made of farmyard local chicken, spicy little chili, fragrant big chili, homemade special chili bean paste and Sichuan people's favorite pickled pepper and ginger by stir-frying, braising and stewing. It has been rated as Ziyang Famous Dish. Furthermore, this store makes three dishes with one chicken, namely, Stir-fried Chicken Giblets with Pickled Pepper, Braised Chicken Offal with Vermicelli, and Chicken Blood in Sour Soup, which are delicious in taste, endless in aftertaste, and loved by the customers.

小贝壳酒店 | LITTLE SHELL HOTEL 19

小贝壳酒店位于乐至县天池镇，1996年创立。经过23年的匠心经营，已拥有中餐营业面积1万余平方米，包括30个包间，住宿、棋牌4 600余平方米，年接待食客达90余万人次，成为乐至县境内独具魅力的园林式川菜专业酒店。2012年被评为四川餐饮名店，2017年荣获川菜辉煌30年卓越企业奖。

Little Shell Hotel is located in Tianchi Town, Lezhi County. It was founded in 1996. After 23 years of development, it has a Chinese food business area of more than 10,000 square meters, including 30 private rooms, and more than 4,600 square meters for accommodation, chess and cards entertainment. It has an annual reception of more than 900,000 diners, becoming a garden style Sichuan cuisine professional hotel with unique charm in Lezhi County. In 2012, it was rated as Sichuan Famous Restaurant, and in 2017, it won Sichuan Cuisine Brilliant 30 Years Outstanding Enterprise Award.

该酒店拥有技术力量较强的厨师团队，注重川菜的传承和创新。1996年首创的盘龙鳝、百年坛子肉、绿毛竹烤肉，受到食客的普遍喜爱。先后研发、推出的多款特色浓郁的菜肴，如观音浸白肉、无量醪糟红烧肉、极品鲜虾粥、旭日东升、小贝壳倒罐蒸桑叶土鸡、鲟鱼汤等，单品年销售量达万份以上。其中，极品鲜虾粥和脆皮粉蒸肉被评为省级川菜名菜，倒罐蒸桑叶土鸡在四川省第二届地方旅游特色菜大赛上荣获传统菜金奖。

The hotel has a strong technical team of chefs, focusing on the heritage and innovation of Sichuan cuisine. In 1996, the pioneering Reclining Dragon Finless Eel, Century-old Tanzi Pork, and Grilled Pork in Green Bamboo Tubes are popular with diners. It has developed and launched a number of dishes with strong characteristics, such as Pork Rolls with the Vegetable, Wuliang Brown Braised Pork with Fermented Glutinous Rice, Excellent Shrimp Porridge, Rising Sun, Little Shell Poured Pot Steamed Mulberry Leaf Local Chicken, Sturgeon Soup, etc.; the annual sales volume of a single dish is more than 10,000. Among them, Excellent Shrimp Porridge and Steamed Pork with Crispy Skin were rated as Provincial Famous Sichuan Cuisine, and Chicken Steamed with Mulberry Leaves won Gold Medal for Traditional Dishes in Sichuan Second Local Tourism Specialty Competition.

玲路酒店 | LINGLU HOTEL 20

玲路酒店位于乐至县帅乡大道800号，于2013年开业，是乐至县首家五星级综合性酒店。

Located at No. 800 Shuaixiang Avenue, Lezhi County, Linglu Hotel was opened in 2013 and is the first five-star comprehensive hotel in Lezhi.

该酒店主体建筑高28层，地下2层，建筑面积达42 000多平方米，外观宏伟壮丽，内部装饰豪华典雅，共有349间豪华舒适的客房及套房，拥有中餐厅、西餐厅、茶室和4个不同规格的会议室、1个多功能会议室及大型购物超市等，能够满足顾客宴请、会务等高端消费需求。酒店全覆盖无线网络等系统，并采用新型节能灯，旨在诠释低碳、环保、节能的主题。在餐饮方面，以"融合、创新"的思想，以川菜为本，集各地著名菜系之所长，创新出独具特色的融合菜品，其招牌菜点有生煎香葱饼、烧汁鲈鱼等。

The main building of the hotel has 28 floors on the ground and 2 floors underground, and the construction area is more than 4, 2000 square meters. The hotel is magnificent outside and luxurious and elegant inside, having a total of 349 luxurious and comfortable guest rooms and suites, Chinese restaurant, western restaurant, tea house, and 4 conference rooms of different sizes, 1 multifunctional meeting room and a large supermarkets, which can satisfy banquets, meetings and other high-end consumer demand. The hotel is fully covered with wireless network and other systems, and adopts new energy-saving lamps to interpret the theme of low carbon, environmental protection and energy saving. Regarding to the catering, with the idea of "fusion, innovation", based on Sichuan cuisine, and gathered the advantages of famous regional cuisine, it has created distinctive fusion dishes, and the signature dishes include Pan-Fried Chive Cake, Braised Bass with Sauce, etc.

21 CHUAN FAMILY RESTAURANT 川家人酒楼

川家人酒楼位于乐至县帅乡大道726号，其前身为乐至县老年活动中心酒楼，于2001年开业，被评为乐至县餐饮十大名店。

Chuan Family Restaurant is located at No. 726, Shuaixiang Avenue, Lezhi County, formerly known as Lezhi Senior Citizens Activity Center Restaurant. It was opened in 2001 and was rated as "Top 10 Famous Restaurants" in Lezhi.

该酒楼是以经营川菜为主的中餐馆，倡导生态、绿色、健康、环保的餐饮理念，承袭传统川菜制作技艺，同时博采众长，菜肴品种多样，口味多变，不断满足人们的口味需求，制作经营的经典菜肴有川家牛排、黄焖土鸡、酱肉大包、苦瓜老鸭汤等。其中，川家牛排被评为乐至县十大名菜。该酒楼还围绕乐至县蚕桑养殖特色，创新推出了桑叶宴，以桑叶为主要食材，创制出凉拌桑叶、翡翠鲫鱼、桑叶炖土鸡等系列菜点品种。

乐至县城鸟瞰
A Bird`s-Eye View of Lezhi County

The restaurant is a Chinese restaurant specializing in Sichuan cuisine, advocating the catering concepts of ecology, green, health, and environmental protection. Following traditional cooking skills of Sichuan cuisine, and learning widely from others' strong points, it creates a variety of dishes with varied tastes, constantly meeting customers' needs. The classic dishes produced are Sichuan Native Steak, Braised Local Chicken, Big Steamed Stuffed Bun with Spiced Pork, Duck Soup with Bitter Melon, etc. Among them, Sichuan Native Steak is rated as "Top 10 Famous Dishes" in Lezhi. Focusing on the characteristics of sericulture of Lezhi, the restaurant brings forth the Mulberry Leaf Feast, which takes mulberry leaves as the main material, creating a series of dishes such as Cold Mulberry Leaves Salad, Emerald Crucian, Stewed Local Chicken with Mulberry Leaves, etc.

乐至壹号烧烤店　LEZHI NO. 1 PORK BBQ RESTAURANT　22

乐至壹号烧烤店位于乐至县天池镇千业路3号，于2015年开业，曾获乐至第八届美食烧烤节十大人气美食烧烤、资阳好味道一等奖和乐至十大名店等称号。

Located at No. 3, Qianye Road, Tianchi Town, Lezhi County, Lezhi No. 1 Pork BBQ Restaurant was opened in 2015 and has won the titles of Top 10 Popular BBQ of the 8th Food and Barbecue Festival, First Prize of Ziyang Good Taste in ziyang and Top 10 Famous Restaurants in Lezhi.

该店的装修风格颇具小资情调，店面较小，主要经营乐至烧烤这一特色美食，荤素皆备，品种丰富，味道佳美，深受消费者喜爱，食客盈门。2018年，乐至壹号烧烤店的升级店——"匠派烧烤"在乐至县天池镇千业路113号开业，成为乐至烧烤特色餐饮的旗舰总店和乐至特色烧烤的升级示范店。该店营业面积1 300

余平方米，可同时容纳500人聚餐，最鲜明的特色是将烧烤美食与歌舞融合在一起，凡是在该店用餐，即可免费K歌，其商务包间中都可以边用餐边唱KTV，而且在用餐大厅，每晚都有歌手驻唱与歌舞表演。此外，该店装修充满动漫元素，风格时尚，每个包间从名称到装饰各不相同，深得年轻消费群体的青睐。

The decoration of the small store is petty stylish, and it is mainly engaged in the specialty of Lezhi barbecue. Meat and vegetables here are available and rich in varieties; the taste is good; the barbecue is loved by consumers; and the diners are packed. In 2018, "Jiangpai Barbecue", the upgrade store of Lezhi No. 1 Pork BBQ Restaurant opened at No. 113, Qianye Road, Tainchi Town, Lezhi County. It has become the flagship headquarters of Lezhi barbecue featured catering and the upgraded demonstration store of Lezhi special barbecue. The business area of the new store is more than 1,300 square meters, which can accommodate more than 500 people dining at the same time. The most distinctive feature here is the fusion of delicious barbecue and song and dance. Anyone who eats in this restaurant can enjoy free KTV. In business private rooms, customers can eat while singing KTV; and in the dining hall, there are singing and dance performances every night. In addition, the store is decorated with animation elements and fashionable style. Each private room is different from name to decoration. It is deeply favored by young consumer groups.

23 EIGHT RICE GRAINS LOCAL DISHES | 八颗米土菜馆

八颗米土菜馆位于乐至县文庙街258号，创立于2014年，是乐至县乡土特色浓郁的餐饮企业。

Located at No. 258, Wenmiao Street, Lezhi County, Eight Rice Grains Local Dishes was founded in 2014. It is a catering enterprise with strong local characteristics in Lezhi.

该菜馆的装修以乡土风格为主，青砖、木门、实木桌凳、灯笼、土碗等是标配，以接待散客零餐为主，可承接婚宴、寿宴等各类宴席，营业面积1 300平方米，拥有2个宴会大厅和10余个豪华包间，可同时容纳600人就餐。该菜馆提倡"光盘行动"，以"真心出美味，真诚遇顾客"为经营理念，以家的感受、细致服务、公道的价位赢得民众之心，主要制作、经营乡土菜和家常菜，乡土气息浓郁，代表菜点有烤黑山羊腿、紫薯芋角、野菜煎饼等。

The decoration of the restaurant is full of country style. It is standardly equipped with gray bricks, wooden doors, solid wood tables and stools, lanterns, soil bowls, etc. The restaurant mainly receives individual guests and meals, and can undertake wedding banquet, birthday banquet and other kinds of banquet. The business area of the restaurant is 1,300 square meters with 2 banquet halls and more than 10 luxury private rooms, which can accommodate 600 people dining at the same time. The restaurant advocates Clean Plate Campaign. With the business philosophy of "cooking sincerely, treating faithfully", the restaurant attracts customers by home feeling, meticulous service, and fair price. It mainly produces and serves local dishes and home-cooked dishes, which have rich country flavor. Its representative dishes include Roasted Black Goat Leg, Purple Potato Taro Cake, Wild Vegetable Pancake, etc.

24 GENERAL STORE OF HECHI WANG'S SPICY NOODLES OF LEZHI | 乐至鹤池王辣面总店

乐至鹤池王辣面总店位于乐至县金穗路92号，于2000年开业，被评为资阳名小吃，2011年获得国家商标局注册。

Located at No. 92, Jinsui Road, Lezhi County, the General Store of Hechi Wang's Spicy Noodles was opened in 2000 and was rated as Ziyang Famous Snack. It was registered by State Trademark Bureau in 2011.

该店店面较小，专营特色面食，制作经营的辣面、烧肉面、豌豆杂酱面、肥肠面、牛肉面等风味独特，麻辣芳香，清淡爽口，柔韧劲道，并且营养丰富，深受食客的喜爱，成为乐至面食店最具代表性的品种。如今，乐至鹤池王辣面不仅有一家总店，还有7家分店，连锁经营取得了一定成效。

The store is small, specializing in specialty food made of flour, such as Spicy Noodles, Braised Pork Noodles, Noodles with Fried Pea and Meat Sauce, Pork Intestines Noodles, Beef Noodles, etc. The noodles of the restaurant have unique flavor, spicy aroma, light, refreshing, smooth and pliable taste, and rich nutrition. The noodles are deeply loved by diners and have become the most representative variety of Lezhi noodle store. Today, there is not only one general store, but also seven branches. The chain operation has achieved some success.

Chapter Five

CELEBRITIES AND ALLUSIONS OF ZIYANG CUISINE

Since 35,000 years ago the ancient "Ziyang Man" started the history of human civilization in Sichuan, there are many sages and celebrities coming out of Ziyang, including Chang Hong, the teacher of Confucius, Wang Bao, litterateur of the Han Dynasty, Dong Jun, Confucianist of the Han Dynasty, Jia Dao, poet of the Tang Dynasty and Chen Tuan, neo-Confucianist of the Song Dynasty, Qin Jiushao, mathematician of the Song Dynasty, Marshal Chen Yi, proletarian revolutionist, militarist, diplomat and poet, and famous writers and calligraphers and painters Shao Zinan, Zhou Keqin, Liu Xinwu, Xie Wuliang etc., as well as many cooking masters of Ziyang cuisine, entrepreneurs, cultural celebrities, and so on. "In the land of abundance, people are numerous and products are plentiful." In their life stories, we can more or less find wonderful moments that link them with various aspects of food.

第五篇 资阳

美食名人与典故

自从三万五千年前古老的『资阳人』开启了四川人类文明史后，资阳这块土地上先后走出了众多的圣贤与名人，有孔子之师苌弘、汉代文学家王褒和经学家董钧、唐代诗人贾岛，以及宋代理学家陈抟、数学家秦九韶，有无产阶级革命家、军事家、外交家、诗人陈毅元帅，有著名作家和书画家邵子南、周克芹、刘心武、谢无量等，还有许多资阳美食的烹饪大师、企业家、文化名人，等等。『天府之国，物阜民丰』，在他们的人生故事里，我们大多能寻觅到他们与美食方方面面之间的精彩片段。

ANCIENT CELEBRITIES AND FOOD ALLUSIONS
古代名人与美食典故

壹 I

苌弘与资阳名馔苌弘鲶鱼 | CHANG HONG AND THE FAMOUS ZIYANG CUISINE OF CHANGHONG CATFISH — 1

  苌弘（约公元前582年~公元前492年），字叔，资阳忠义镇高岩山人，是春秋末期著名的天文学家、音乐大师和阴阳家，资阳古代三贤之首。苌弘自幼聪明，博学多才，擅长天文，精通音律，著有《海内经》《大荒东经》15篇，享有"智多星"的美名。"孔子问于乐"，说的就是孔子与苌弘交往的故事。公元前521年，孔子不远千里，专程拜苌弘为师，请教音乐理论和天文知识，交流治理天下的思想，二人结下了深厚情谊。三个月后，孔子满意而归，后来写出了《乐经》等不朽名著。而苌弘因其智慧超群、精通天文历律和阴阳学被东周三代君王重用，致力于东周王室的巩固，苦苦支撑风雨飘摇的周王朝。公元前492年7月，周敬王偏听偏信，中了晋国贵族的离间计，用酷刑杀害了年近九旬的苌弘。苌弘临死前沉痛呼喊："杀身之祸我并不悲哀，我痛惜的是宗周不统一！"三代辅国之臣为国冤死，河南禹县乡民们将苌弘的鲜血藏在一个匣中，三年后变成一块璀璨夺目、光彩照人的碧玉。由此，成语中有了"碧血丹心"一词，晋代左思《蜀都赋》有了"碧出苌弘之血"的千古名句。

  Chang Hong (around 582 B. C.—492 B. C.), whose courtesy name was Shu, was born in Ganyan Mountain, Zhongyi Town, Ziyang City. He was a famous astronomer, musician and Yin-Yang specialist in the late Spring and Autumn Period, and was the first of the ancient three sages in Ziyang. Chang Hong was clever and knowledgeable when he was young, and he was good at astronomy and temperament. He wrote 15 articles and books such as *Hainei Jing* and *Dahuangdong Jing*, and enjoyed the reputation as "mastermind". "Confucius Asking about Music" referred to the story of Confucius and Chang Hong. In 521 B. C., Confucius traveled thousands of miles to learn music theory and astronomy knowledge from Chang Hong, and to exchange ideas about governing the world. They formed a deep friendship. Three months later, Confucius returned with satisfaction, and wrote *Classic of Music* and other immortal classics afterwards. Because of his superior wisdom, knowledge of astronomical calendar and Yin-Yang

Theory, Chang Hong was put in important positions by the three kings of the Eastern Zhou Dynasty. He devoted himself to the consolidation of the Eastern Zhou Dynasty and struggled to sustain the shaky Zhou Dynasty. In July, 492 B. C., King Jing of Zhou listened selectively and fell for the trick of nobility of Kingdom Jin, and killed Chang Hong, who was almost 90 years old, with torture. Chang Hong shouted painfully before he died, "I am not sad about my death, but I deeply regret the disunity of the Zhou Dynasty!" The official assisting the three generations of the kingdom died of persecution, thus in honor of Chang Hong, people of Yu County in Henan Province hid his blood in a small box. Three years later, the blood turned into dazzling and brilliant jade. Therefore, the phrase "bi xue dan xin" (means righteous blood and loyal heart) came out; Zuo Si of the Jin Dynasty wrote in his *Shudu Fu* the famous saying "Jade comes out of Chang Hong's blood".

苌弘的生平故事在其家乡民间广为传颂，人们为了纪念这位忠义良臣，将苌弘故乡附近的山水命名为苌弘山、苌弘洞、苌弘桥、苌弘村等。在资阳地区，人们也因对苌弘的拥戴与景仰，加之民风素来热情友善，精于制作美馔佳肴，于是为孔子访苌弘这段历史佳话中增添了美食的记忆。相传孔子拜访苌弘时，二人相见甚欢，苌弘用家乡沱江中出产的鲶鱼，亲自为孔子烹制鲶鱼美馔，于是，"苌弘鲶鱼"便流传开来。改革开放以后，资阳的餐饮业者更对苌弘鲶鱼进行文化挖掘与工艺提升，此菜曾获评资阳市"十二大经典名菜"之首。如今，苌弘鲶鱼烹调技艺被列入资阳市非物质文化遗产代表性名录，苌弘鲶鱼又获得了资阳市著名川菜、畅销特色菜等多项荣誉称号，这是源于其精湛的烹调技艺和独特风味，也是源于其独特的文化内涵，是资阳人民铭记苌弘与孔子两位先哲的美好方式。

The life stories of Chang Hong were widely spread in his hometown, and in memory of this loyal official, people named the landscape around Chang Hong's hometown as Changhong Mountain, Changhong Cave, Changhong Bridge and Changhong Village, etc. In Ziyang area, people loved and admired Chang Hong, and since they were warm, friendly and good at making delicious food, they added food memories to the historical story of Confucius visiting Chang Hong. Legend has it that when Confucius visited Chang Hong, they had a good time and Chang Hong cooked catfish for Confucius himself with the catfish grown in the Tuojiang River in his hometown. Since then, "Changhong Catfish" spread gradually. After the Reform and Opening Up, people in the catering industry in Ziyang explored the culture and improved the craft of Changhong Catfish, and this dish was once awarded as top one of the "Twelve Classic Dishes" in Ziyang. Today, the cooking skills of Changhong Catfish have been included in the representative list of Ziyang municipal intangible cultural heritage. Besides, the dish has also obtained the titles of Famous Sichuan Cuisine in Ziyang, Popular Specialty, etc. Due to its superb cooking skills, the peculiar flavor, as well as the unique cultural connotation. It is also a special way for people in Ziyang to remember the two masters, Chang Hong and Confucius.

王褒与茶 | WANG BAO AND TEA

  王褒（公元前90年~前51年），字子渊，资阳临江镇墨池村墨池坝人，是西汉时期著名的辞赋家，与扬雄合称"渊云"。王褒虽出生于蜀中山野，自幼参与农耕，但从未放弃读书治学，博览群书，精通六艺，为风度翩翩、名震四方的英俊才子。汉宣帝时，王褒创作了多首辞赋歌颂汉朝盛世，宣帝招他入朝作《圣主得贤臣颂》《甘泉赋》等，得到宣帝喜爱，提升其官职。王褒才华横溢，谈吐风趣，成为汉宣帝时期的文坛领袖，创作的辞赋较多、成就显著，但是，在中国饮食史上最著名、最重要的却是他在年轻时客居成都所写的一篇《僮约》，即买奴隶的契约。该文在无意之间成为了中国茶文化最早的记录。

  Wang Bao (90 B. C.—51 B. C.), whose courtesy name was Zi Yuan, was born in Mochi Ba, Mochi Village, Linjiang Town, Ziyang city. He was a famous poet and writer in the Western Han Dynasty, and was collectively called "Yuan Yun" together with Yang Xiong. Although Wang Bao was born in the mountains and fields of Shu and engaged in farming since his childhood, he never gave up studying. Reading widely and mastering the Six Arts, he was a handsome wit with grace and big reputation. During the Emperor Xuan Period of the Han Dynasty, Wang Bao wrote many poems and prose to praise the prosperous times of the Han Dynasty. Emperor Xuan invited him to the court, and he wrote *Ode to the Wise Lord Getting the Virtuous Officials*, *Ganquan Fu*, etc., which were loved by Emperor Xuan, thus he promoted Wang Bao's official position. With his brilliant intellect and witty speeches, Wang Bao became the literary leader during the Emperor Xuan Period of the Han Dynasty, and he wrote many poems and prose with remarkable achievements. Nonetheless, his most famous and important article in the history of Chinese food was the *Tongyue* written when he lived in Chengdu as a young guest, that is, a contract for slave buying. This article inadvertently became the earliest record of Chinese tea culture.

  公元前59年正月，王褒客居在成都友人家中，友人的遗孀杨氏有一个名为便了的中年家奴，王褒常常支配他去买酒。便了认为王褒是外人，很不情愿为他跑腿办事，有一天到男主人的墓前抱怨说："您当初买我的时候，只是让我看守家里，可没有让我为别的男人买酒啊。"王褒知道了这件事，很生气，就在正月十五这一天花钱从杨氏手中买下便了。便了虽然不情愿，却也没有办法，不过他提出要求："既然如此，您要像杨家买我时那样，把以后我应当做的事在契约中写明白，不然我可不干。"王褒为了惩治便了，就写下了一篇戏谑性的《僮约》作为契约。此文别开生面，字句生动诙谐，其中有这样的描述："脍鱼炮鳖，烹茶尽具"，"武阳买茶"。意思是说，家里来了客人，要烧水烹茶，待客人和主人饮完茶后要洗净茶具，好好收藏；如果没有茶叶了，就要到武阳去买回茶叶。王褒这段关于茶的文字，不仅写了烹茶，还写了珍藏茶具，更写了茶的交易市场，奠定了中国茶文化的基础。由这一记载可以知道，四川地区是全世界最早种茶与饮茶的地区，武阳（今四川省眉山市彭山区）是当时茶叶主产区和著名的茶叶市场，王褒也因此文而成为中国乃至世界第一位记载茶的种植、饮用和茶叶市场之人。

In the lunar January of 59 B. C., Wang Bao lived as a guest in the house of a friend in Chengdu. His friend's widow, Mrs. Yang, had a middle-aged family slave named Bian Liao. Wang Bao often dictated him to buy liquor. Bian Liao considered that Wang Bao was an outsider, and was very reluctant to run errands for him. One day he went to his host's grave and complained, "When you bought me, you only asked me to guard the house, but you didn't ask me to buy liquor for other men." Since knowing this matter, Wang Bao was very angry, and he bought Bian Liao from Mrs. Yang on the fifteenth day of the first lunar month. Although Bian Liao was reluctant, he had no choice, and he asked. "In that case, you should write in the contract what I should do in the future, just as the Yang family did when they bought me, or I will not do it." In order to punish Bian Liao, Wang Bao wrote the playful *Tongyue* as the contract. The article is fresh, vivid and witty, with descriptions of "slicing fresh fish and stewing turtle, boiling tea and cleaning tea set", "buying tea in Wuyang". This means that when a guest comes to the house, he must boil water and prepare tea, and after the guest and the host drink the tea, he must wash the tea set and collect it well; if there is no tea, he has to go to Wuyang to buy tea. Wang Bao's description of tea not only describes the boiling of tea, but also the collection of tea set, and even the trading market of tea, laying the foundation of Chinese tea culture. It can be known from this record that Sichuan was the first region in the world to plant and drink tea, and Wuyang (now Pengshan District, Meishan City, Sichuan Province) was the main producing area and famous tea market at that time. Therefore, Wang Bao became the first person in China and even in the world to record the cultivation and drinking of tea, and the tea market.

3 DONG JUN AND THE DIET ETIQUETTE OF THE HAN DYNASTY 董钧与汉代食礼

董钧（约公元前13年~62年），字文伯，资阳东安乡金山寺村桐梓坝人，汉代著名经学家、教育家。他为汉朝制订礼仪制度，也包括汉朝的饮食礼仪与制度，对后世产生了深远的影响。

Dong Jun (about 13 B. C.—62 A. D.), whose courtesy name was Wen Bo, was born in Tongzi Ba, Jinshansi Village, Dong'an Township, Ziyang City. He was a famous Confucianist and educator in the Han Dynasty. He made the etiquette system for the Han Dynasty, including the diet etiquette and system of the Han Dynasty, which had a far-reaching impact on later generations.

董钧精通儒家经学，16岁时就被推荐进入皇宫，成为掌管宗庙祭祀时祭品准备的官员。东汉初年，董钧被委任为大司徒府郎官的职务，着力研究礼仪，对古代礼法、礼义作了权威的研究、记载和解释，并且著书立说，成为后世称道的汉代礼仪制度的制定人。董钧也曾被任命为太常博士，管理宫廷礼仪，教育皇亲国戚的子弟。《后汉书·董钧传》称董钧"平生以授徒讲学为乐，常教授门生百余人，当世称为通儒"。董钧对于传播儒家思想、发展汉代礼学发挥了积极作用，对后世产生了深远的影响。"夫礼之初，始诸饮食"，董钧也十分重视饮食礼仪的制定与传播，为汉代宗庙祭祀的献食制度、宫廷宴乐制度等的制定作出了开创性的贡献。他一生受皇室重用，年逾七十后才回到故乡资阳，病逝于桐梓坝。董钧作为博学多才的学者、汉代礼制礼学的制定者和积极推行者，对汉代饮食礼仪制度的制定和推行起到了重要作用。

Proficient in Confucian classics, Dong Jun was recommended to the imperial palace at the age of 16 as an official in charge of the preparation of sacrificial offerings for the jongmyo fete. In the early years of the Eastern Han Dynasty, Dong Jun was appointed as the Langguan of Dasitu Office. Then he focused on the study of etiquette, made authoritative research, record and interpretation of ancient rites and courtesies, and wrote relevant books, becoming the praiseworthy founder of the etiquette system of the Han Dynasty. Dong Jun was also appointed as Taichang Boshi, who managed court etiquette and educated the children of the royal relatives. In *Later Han · Biography of Dong Jun*, it is recorded that Dong Jun "enjoyed teaching in his life, usually had more than 100 students and was called expert Confucian scholar." Dong Jun played an active role in spreading Confucianism and developing theory of rituals in the Han Dynasty, which had a profound influence to the later generations. "Diet etiquette is the origin of human etiquette." Dong Jun also attached great importance to the establishment and dissemination of diet etiquette, and made pioneering contributions to the establishment of the offering food system in the jongmyo fete, the palace banquet music system and other systems of the Han Dynasty. He was appointed important positions by the emperor in his life, and returned to the hometown Ziyang after he was over 70, and died of illness in Tongzi Ba. As a knowledgeable scholar, and the maker and active promoter of the system of rites and theory of rituals in the Han Dynasty, Dong Jun played a significant role in the establishment and implementation of the diet etiquette system in the Han Dynasty.

4 聂氏父子、陈兴友与临江寺豆瓣
NIE'S FATHER AND SON, CHEN XINGYOU AND LINJIANGSI CHILI BEAN PASTE

清康熙二十九年（公元1691年），江西人聂长松举家从丰城迁徙到四川，定居资阳临江乡凉水村何家湾，即今天之临江寺豆瓣厂所在地，聂长松、聂志兴父子以事农和贩卖酱油、食醋与豆瓣为生。但贩运道路艰辛、利润微薄，聂志兴便准备自办酱园，他让儿子聂守荣去别家酱园学艺，买地种蚕豆、豌豆和辣椒。1737年，聂家找到并买下了唐代武则天赐建蒙刺寺遗留的"菩提""伽叶"两口古井；1738年，聂家创办临江寺第一家酱园——义兴荣酱园，取古井水酿制豆瓣、酱油和食醋等调味品。1740年，祖籍资州的三朝御厨陈兴友告老还乡，聂志兴喜出望外，携其子聂守荣登门向陈兴友求教，却遭婉言谢绝。聂家父子并未气馁，聂守荣天天上门为其砍柴、扫地、担水，长达半年，陈兴友深受感动，遂将宫廷豆瓣的制作秘方传给聂守荣，并辅佐指导聂守荣经营酱园，成就了聂氏豆瓣咸、甜、辣、香、鲜的风味特征。聂家豆瓣原料考究，鲜红辣椒采用养马河出产的，色鲜肉厚，蚕豆采用罗汉沱或临江寺出产的，柔和化渣，仔姜要用七里坪出产的，并且必须采自白露前，以确保鲜、嫩、脆。1781年，相传乾隆皇帝70岁寿诞，紫禁城举办千叟宴，宣诏年长的昔日近臣入宫贺寿，时年84岁的陈兴友也在列，陈兴友入宫时将自己精心指点制作的聂氏豆瓣贡奉皇帝，乾隆品尝了用此豆瓣烹饪的菜品后觉得滋味上佳，龙颜大悦，遂赐名为"八宝豆瓣"。

In the 29th year of Kangxi Period of the Qing Dynasty (1691), Nie Changsong, a native of Jiangxi Province, moved his family from Fengcheng to Sichuan and settled in Hejiagwan, Liangshui Village, Linjiang Township, Ziyang, which is the location of Linjiangsi Temple Bean Paste Factory today. Nie Changsong and his son Nie Zhixing lived on farming and selling soy sauce, vinegar and bean paste. However, the roads of transporting and trading goods were difficult and the profit was meager. Nie zhixing planned to set up his own sauce and pickle factory. He sent his son Nie Shourong to another sauce and pickle factory to learn skills and bought land to grow broad beans, peas and chilies. In 1737, Nie's found and bought two ancient wells of "Puti" and "Jiashe" which were given titles by Empress Wu Zetian of the Tang Dynasty when build the Menglasi Temple. In 1738, Nie's set up the first sauce and pickle factory in Linjiangsi Temple—Yixinrong Sauce and Pickle Factory, which brewed bean paste, soy sauce, vinegar and other condiments with water from the ancient wells. In 1740, Chen Xingyou, the royal cook of three dynasties and whose ancestral home was Zizhou, retired and returned to his hometown. Nie Zhixing was overjoyed. He and his son Nie Shourong visited Chen Xingyou for advice, but were politely refused. Nie's father and son were not discouraged. Nie Shourong visited every day to cut firewood, clean the floor and carry water for him for half a year. Chen Xingyou was deeply moved, so he passed the secret recipe of palace bean paste to Nie Shourong, and assisted and guided Nie Shourong to run the sauce and pickle factory, achieving the salty, sweet, spicy, fragrant and fresh flavor characteristics of Nie's bean paste. The raw materials of Nie's bean paste were exquisite; fresh chilies used were from in Yangmahe with bright color and thick flesh, broad beans were from Luohantuo or Linjiangsi Temple with soft taste and an easy-to-melt quality, tender gingers were from Qiliping and must be picked before White Dew to ensure the fresh and

crisp flavors. It is said that, in 1781, during Emperor Qianlong's 70th Birthday, the Forbidden City held Qiansou Feast, summoned elder former close officials to the palace to celebrate his birthday. Chen Xingyou was 84 years old at that time and was also among them. He contributed the Nie's bean paste when he went to the palace, which was produced under his elaborate instruction. After tasting the dishes cooked with this bean paste, Emperor Qianlong felt they were delicious and was very pleased, hence gave the name of "Eight Treasures Bean Paste".

由于何家湾山清水秀、乔木耸立、绿竹环绕、水质极佳，沟内地势开阔平坦，日照充足，更加上聂家豆瓣的声名鹊起，引得众多的酱园纷纷开张，当时的民谣称"一过姑嫂坳，就闻豆瓣香"。整个临江寺已成为制作、贩运豆瓣的一条街，车水马龙，码头商贾云集，盛况空前，资州四大门、成都东大街及其他州府闹市均设有临江寺豆瓣专卖门店。进入19世纪，聂氏后代分家立业，先后分为义兴祥、义兴福和义兴荣三家。其中，义兴荣于1930年更名为中兴祥，在聂季烈主管时期十分兴盛，被称为"豆瓣世家"。1931年，成渝公路全线贯通，临江寺豆瓣销量大增。1937年，各家酱园蓬勃发展，最高年产量达10万公斤。1954年，公私合营，当地5家酱园合办资阳县临江乡酱园联营社，仍由聂季烈经营。临江寺豆瓣深受毛泽东、朱德、贺龙、邓小平等党和国家领导人的喜爱和赞赏，因此也成为享誉中南海的资阳调味品。

Because Hejiawan had beautiful mountains and clear waters, tall trees and green bamboos, the water quality was excellent, the terrain inside was open and flat, the sunshine was abundant, and Nie's bean paste rose to fame, which attracted many sauce and pickle factories to open. Ballads of that time said, "Once pass Gusao Ao, one can smell the fragrance of the bean paste." the whole Linjiang Temple region gradually became a street for producing and selling bean paste, with heavy traffic and numerous merchants gathering in wharf unprecedented pomp. Linjiangsi Temple bean paste exclusive stores were set up at the four major gates of Zizhou, on Dongdajie Street of Chengdu and in downtowns of other states and cities. In the 19th century, Nie's descendants separated their business into three parts, namely, Yixingxiang, Yixingfu and Yixingrong. Among them, Yixingrong was renamed Zhongxingxiang in 1930, which was very prosperous when Nie Jilie was in charge, and was known as "Family of Bean Paste". In 1931, the Chengdu-Chongqing Highway was completed, and the sales of Linjiangsi Temple bean paste increased greatly. In 1937, sauce and pickle factories were thriving with a maximum annual output of 100,000 kilograms. In 1954, five local sauce and pickle factories jointly established the Sauce and Pickle Affiliated Society in Linjiang Township, Ziyang County, and Nie Jilie was still in charge. Linjiangsi chili bean paste was deeply loved and appreciated by Mao Zedong, Zhu De, He Long, Deng Xiaoping and other Party and state leaders, so it became a famous condiment in Zhongnanhai.

MODERN CELEBRITIES AND FOOD ALLUSIONS
近代至今名人与美食典故

贰 II

赖源鑫与赖汤圆　LAI YUANXIN AND LAI'S TANGYUAN　1

赖源鑫（生卒年不详），资阳东峰镇人，四川名小吃赖汤圆创始人。他在成都制作售卖赖汤圆，苦心经营，发家致富，却仍然不忘家乡，在家乡捐款办学，培养人才。

Lai Yuanxin (the year of birth and death is unknown), born in Dongfeng Town of Ziyang, is the founder of Sichuan famous snack Lai's Tangyuan. He made and sold Lai's Tangyuan in Chengdu. With painstaking management, he built up the family fortune, but still did not forget his hometown. He donated money to run schools and cultivate talents in his hometown.

清代光绪年间，赖源鑫随堂兄从资阳到成都一家饮食店打工、学习厨艺，不久被辞退。为了谋生，他只好向堂兄借了一些钱，置办了一副担子，开始挑起担子，做起汤圆生意。因为是小本经营，要想在成都立足，他给自己立下4条规矩：一是利看薄点，二是周转快点，三是服务好点，四是质量高点。他起早贪黑，走街串巷，早上卖的钱又去备办卖夜汤圆的原料。他以质量取胜，货真价实，精选糯米磨成浆，制成湿米粉，包入糖油重的馅心制成。他制作的汤圆煮时不烂皮、不露馅、不浑汤，吃时不粘筷、不粘牙、不腻口，还外加一碟白糖芝麻酱，色泽洁白，味道香甜，口感爽滑软糯，成为当时成都极负盛名的小吃，被成都人亲切地称为"赖汤圆"。赖源鑫苦心经营数十载，直到1937年才在总府街买下一间店面，开始坐店经营，生意也越做越旺，后来又开起米粮铺和钱庄，逐渐成为在成都的资阳同乡会里举足轻重的人物。他本人目不识丁，却深知知识的重要性，发家之后亦不忘乡梓，得知家乡要筹建一所中学，便捐赠了150担（约合2.5万多公斤）谷子，作为储彦中学的办学经费，此后直到1950年间仍然时有捐助。赖源鑫捐款办学之事，在资阳一直传为佳话。

During the Guangxu Period of the Qing Dynasty, Lai Yuanxin went to a restaurant in Chengdu from Ziyang with his cousin to work and learn cooking. He was soon fired. In order to make a living, he had to borrow some money from his cousin. He bought a carrying pole, and started the business of sweet rice dumpling with the carrying pole. Because it was a shoestring business, he set four rules to himself in order to gain a foothold in Chengdu. The first was to get small profits, the second was to speed up the capital circulation, the third was to supply good service, and the last was to make high quality food. Therefore, he worked from dawn to night and went from street to street. The money he earned in the morning was used to prepare the raw materials of the sweet rice dumplings sold in the evening. He won with the quality, the genuine goods and the fair price. He carefully selected glutinous rice to grind into pulp, then made into wet rice flour, and finally stuffed with fillings heavy with sugar and oil to make sweet rice dumplings. When cooking, the sweet rice dumplings he made were not easy to broken, the stuffing would not leak and the soup was clear. When eating, the sweet rice dumplings did not stick to chopsticks and teeth, and were not greasy. Besides, a plate of sugar sesame paste was added. The sweet rice dumpling was white in color, sweet in taste and smooth and soft in flavor, became a famous snack in Chengdu at that time and was affectionately known as "Lai's Tangyuan". Lai Yuanxin painstakingly managed for decades. Until 1937, he bought a storefront in Zongfu Street and began to sit to operate in a store. His business became more and more prosperous, and also opened the grain shop and the ancient private bank. Gradually, he turned to be a pivotal figure in the Ziyang Association of Chengdu. Though he himself was illiterate, he knew the importance of knowledge. After getting rich, he also did not forget the hometown. Knowing that a middle school was to be built in his hometown, he donated 150 dan (more than 25,000 kilograms) of millet as the funds for running Chuyan Middle School, and the donation was lasted from time to time until 1950. Lai Yuanxin's donation for running a school has always been widely reported in Ziyang.

赖汤圆从诞生至今，已走过百余年的历程，深受四川乃至国内外美食爱好者的追捧，曾有诗云："吃饭最好年夜饭，看戏爱看大团圆，照相要照合家欢，小吃必食赖汤圆！"2006年，赖汤圆被国家商务部认定为第一批中华老字号。2011年，其制作技艺被列入第三批四川省非物质文化遗产代表性名录。一百多年前，从资阳走出的赖源鑫，在成都以自己勤劳的双手演绎了一个美食励志的传奇故事，留下了一个百年金字招牌——赖汤圆。

It has been more than 100 years since the birth of Lai's Tangyuan. It has always been popular with food lovers in Sichuan and even at home and abroad. There was a poem, "the best meal is the New Year's Eve dinner, the best ending is the happy reunion, the best photo is the family portrait, and the best snack is the Lai's Tangyuan!" In 2006, it was recognized by the Ministry of Commerce as one of the first batch of Chinese Time-Honored Brands. In 2011, its production techniques were included in the third batch of the representative list of Sichuan provincial intangible cultural heritage. Lai Yuanxin, who came out of Ziyang over 100 years ago, performed an inspirational legend about food with his hard-working hands in Chengdu, leaving a century-old golden signboard—Lai's Tangyuan.

2 CHEN MINGZHAI AND WEIXINZHAI SAUCE AND PICKLE FACTORY 陈铭斋与味心斋酱园

陈铭斋（1898年~1939年），资阳广福场人，后迁居金堂县赵镇，于20世纪30年代在当地创办味心斋酱园，将临江寺豆瓣制作工艺改良并发扬光大，销售到各地。

Chen Mingzhai (1898—1939) was a native of Guangfu Field of Ziyang, who later moved to Zhaozhen Town, Jintang County. In the 1930s, he founded Weixinzhai Sauce and Pickle Factory there, which improved and carried forward the production technology of Linjiangsi chili bean paste and sold it all over the country.

1930年，陈铭斋从资阳的临江寺聘请熟练酿造工人做味心斋酱园掌缸师，生产的豆瓣质量上乘，瓣大化渣、柔中带脆，并制作出金钩、香油、火腿、鱼松、墨鱼等不同味型的豆瓣，适用于佐餐、烹菜。其后，他又通过试验，制作了耐压不怕碰损、便于携带的瓶装豆瓣，瓶外涂刷紫红油漆、彩印商标，以油纸和彩纸封口、红麻绳扎提，色泽鲜艳。味心斋豆瓣远销陕西、甘肃等地，抗日将士出征时也买来做行军的下饭菜，各地商人也将其作为礼品带到上海、香港等地。1935年，陈铭斋在赵镇捐资、兴办儿童习艺所。每到过年，他还向当地贫民捐米捐粮。陈铭斋味心斋酱园的成功，既显示出临江寺豆瓣作为川中知名调味品的重要影响力，更彰显了资阳人的勤劳与智慧。

In 1930, Chen Mingzhai hired skilled brewers from Linjiangsi Temple in Ziyang as the masters in charge of sauce jars of Weixinzhai Sauce and Pickle Factory. The bean paste produced was of superior quality, with big flaps, soft and crisp taste. There were bean pastes of different flavors, such as dried shrimp, sesame oil, ham, fish floss, cuttlefish and so on, which were suitable for eating and cooking. Thereafter, through trials and errors, he made bottled bean pastes which were pressureproof, damage-proof, and easy to carry. Being coated with purple and red paint and color trademark, sealed with oil paper and color paper, and tied with red hemp rope, the bottles were bright in color. Weixinzhai bean paste were sold in Shaanxi, Gansu and other places. The anti-Japanese soldiers also bought them before going out to war as the dishes eaten together with rice on the march, and merchants of different regions brought them to Shanghai and Hong Kong and other places as gifts. In 1935, Chen Mingzhai donated money and set up a children's art institute in Zhaozhen Town. On every Chinese New Year, he also donated rice and grain to the local poor. The success of Chen Mingzhai's Weixinzhai Sauce and Pickle Factory not only shows the important influence of Linjiangsi chili bean paste as a famous condiment in Sichuan, but also manifests the diligence and wisdom of Ziyang people.

邹海帆父子与安岳柠檬 | ZOU HAIFAN AND HIS FATHER AND ANYUE LEMON 3

邹海帆（1907年~1969年），资阳市安岳县龙台场人，1928年考入华西协合大学牙学院，1937年毕业后留校任教，我国著名牙周病学家、华西口腔病院第二任院长，也是将柠檬引入中国安岳的第一人。他于1929年将柠檬这种神奇水果引入家乡，其父邹江亭精心培育出安岳柠檬，使柠檬落户安岳。数十年之后，柠檬在安岳大放异彩，造就了"中国柠檬之都"。

Zou Haifan (1907—1969) was a native of Longtai Feild, Anyue County, Ziyang City. He was admitted to the School of Dentistry of West China Union University in 1928. After graduation in 1937, he taught there. He was a famous periodontist in China, the second Dean of West China Hospital of Stomatology, as well as the first person to introduce lemon to Anyue in China. In 1929, he introduced the magic fruit lemon to his hometown. Then his father Zou Jiangting cultivated Anyue lemon meticulously, making lemon settled in Anyue. After several decades, lemon yielded unusually brilliant results in Anyue, and made Anyue "Hometown of Lemon in China".

1928年6月，邹海帆考入华西协合大学牙学院。刚进入学校，他就被校园周边教师别墅院子里的苹果、梨、柠檬、柑橘、李子、柿子、桃子等各种果树深深吸引。邹海帆来自一个中医世家，父亲邹江亭精于医术、擅长园艺，1914年就创办实业果园，大力发展果树种植产业。家庭的影响使邹海帆对校园内这些国内外的优良果树品种产生了浓厚兴趣，心中萌生了要将其中一些四川没有的优良果树引入到自己的家乡安岳县龙台场的念头。在华西协合大学生物系加拿大籍教授丁克生的帮助下，邹海帆找到了适合安岳县土质和气候，从美国加利福尼亚州引入华西的尤力克（阿卡）柠檬。1929年，他将尤力克柠檬接穗，得到柠檬果苗并寄回安岳龙台镇，邹江亭收到儿子寄回的柠檬果苗后精心培育、嫁接发展，经过多

年选优提纯，培养出了适合安岳生态条件的优良株系——安岳柠檬。从此，安岳柠檬成为最受国人喜爱的柠檬品种，邹江亭和安岳县也被赞誉为"安岳柠檬之父"和"中国柠檬之都"。为了纪念邹海帆教授父子对柠檬事业做出的巨大贡献，华西口腔医院内的一片柠檬园被命名为"邹海帆柠檬园"。九十年过去了，今天的安岳县柠檬种植面积已达48万亩，年产量42万吨、占全国的80%以上。"魅力柠海"连绵数十万亩，构成一条壮观的绿色长廊，安岳柠檬也成为中国驰名商标和地理标志保护产品。

　　In June, 1928, Zou Haifan was admitted to the School of Dentistry of West China Union University. As soon as he entered school, he was deeply attracted by apple, pear, lemon, orange, plum, persimmon, peach trees and other fruit trees in the yard of the teacher's villa around the campus. Zou Haifan came from a family of traditional Chinese medicine. His father, Zou Jiangting, was skilled in medicine and good at gardening. He founded industrial orchard in 1914 and vigorously developed fruit tree planting industry. The influence of his family made Zou Haifan great interested in those foreign fine fruit tree varieties on campus, and he initiated the idea of introducing some of these excellent fruit trees which were not available in Sichuan to his hometown Longtai Field in Anyue County. With the help of Ding Kesheng, a Canadian professor in the Department of Biology of West China Union University, Zou Haifan found the Eureka lemon, which was introduced from California of America to West China, was suitable for the soil and climate of Anyue County. In 1929, after the scion of the Eureka lemon, he got the lemon seedling and sent it back to Longtai Town of Anyue. Zou Jiangting carefully cultivated, grafted and developed the lemon seedling sent back by his son. After years of selection and purification, he cultivated an excellent strain—Anyue lemon, which was suitable for the ecological conditions of Anyue. Since then, Anyue lemon has become the most popular lemon variety in China, Zou Jiangting and Anyue County have also been praised as the "Father of Anyue Lemon" and the "Hometown of Lemon in China" respectively. In memory of Professor Zou Haifan's and his father's great contributions to the lemon cause, a lemon garden in West China Hospital of Stomatology was named "Zou Haifan Lemon Garden". Ninety years have passed, today's lemon planting area in Anyue County has reached 480,000 mu, with an annual output of 420,000 tons, accounting for more than 80% of the whole country. "Charismatic Lemon Ocean", which covers hundreds of thousands of mu, constitutes a spectacular green corridor. Anyue lemon has also become China Famous Brand and National Geographical Indication Protection Product in China.

吴仲良与川菜出国　WU ZHONGLIANG AND SICHUAN CUISINE GOING ABROAD　4

　　吴仲良（1921年~2017年），资阳市乐至县龙门乡三星桥人，技艺高超的川菜厨师，资阳地区川菜出国第一人。吴仲良先生更是四川省捐资助学金额最大、时间最长、建学校最多、影响最广的海外侨胞。自1997年起，连续八次在国庆节应邀登上天安门城楼观礼，2002年被四川省侨联聘为名誉主席。

Wu Zhongliang (1921—2017), a native of Sanxing Bridge, Longmen Township, Lezhi County, Ziyang City, is a highly skilled chef of Sichuan cuisine and the first person to go abroad with Sichuan cuisine in Ziyang region. Furthermore, Mr. Wu Zhongliang was one of the overseas Chinese who donated the largest amount of money for education, lasted the longest time, built the most schools and had the widest influence in Sichuan Province. Since 1997, he has been invited to the Tian'anmen Rostrum to watch the ceremony on National Day for eight consecutive times. In 2002, he was appointed Honorary Chairman of Sichuan Overseas Chinese Federation.

　　吴仲良幼年时父母双亡，寄养在舅舅家中，10余岁时为了生计沿街叫卖，后进入餐馆当学徒。学徒期满，他辗转于成都、重庆、南京等地担任川菜馆厨师。后来，他在南京开设了"三六九"川菜馆，该餐馆成为南京新街口最红火的餐厅。1949年，吴仲良前往台湾，仍以开川菜馆为生。1965年，为生计所迫，他在友人的帮助下来到美国，开始了在异国他乡的拼搏人生。吴仲良技艺高超、敬业创新，经他烹调的川菜，既保留了川菜的特色，又迎合了纽约当地食客的要求，因此广受青睐，很快在纽约站稳了脚跟。到1972年，吴仲良已在纽约开办了四川饭店、北京饭店、成渝饭店、重庆饭店共4家川菜餐厅，这些以祖国省、市名字命名的饭店，不仅使旅美的华人、华侨备感亲切，增进了对祖国的感情，也扩大了川菜在美国纽约这个国际大都会的影响力，提升了中国饮食文化在当地食客心中的地位。吴仲良漂泊异国，历尽艰辛，用智慧和汗水浇灌出一朵朵川菜之花，绽放在美利坚。

Wu Zhongliang's parents died when he was young, and was fostered in his uncle's family. When he was over 10 years old, he peddled in the street for a living, and then he entered the restaurant as an apprentice. After his apprenticeship, he worked as chef in Sichuan restaurants in Chengdu, Chongqing, Nanjing and other places. Later, he set up "Three Six Nine" Sichuan restaurant in Nanjing, which became the most popular restaurant in Xinjiekou of Nanjing. In 1949, Wu Zhongliang went to Taiwan, still living on opening Sichuan restaurant. In 1965, forced to make a living, he came to the United States with the help of his friends and began to struggle in a foreign country. Wu Zhongliang was a skilled, dedicated and innovative cook whose Sichuan cuisine, while retaining its characteristics, catered to the requirements of local diners in New York, was so popular that he was soon able to make a living in New York. By 1972, Wu Zhongliang had opened four Sichuan cuisine restaurants in New York, namely, Sichuan Restaurant, Beijing Restaurant, Chengyu Restaurant, and Chongqing Restaurant. These restaurants named after the provinces and cities of the motherland not only made the Chinese and overseas Chinese in the United States feel at home and enhanced the feelings for the motherland, but also enlarged the influence of Sichuan cuisine in the international metropolis of New York and promoted the position of Chinese food culture in the hearts of local diners. Wu Zhongliang led a wandering life in the foreign country, went through all kinds of hardships, and cultivated the flowers of Sichuan cuisine with wisdom and sweat, blooming in the United States.

　　1982年，吴仲良回到故乡资阳扫墓，目睹了乡间学校的简陋、学子辍学，心中难以平静，决心为家乡的教育事业尽一份力量。自1986年起，他把自己的积蓄和部分养老金共计人民币2 000余万元捐赠给家乡兴办教育事业，先后在家乡乐至县和资阳市、内江市多地修建了侨心学校数十所、幼儿园1所、敬老院

1所，设立了金额为350余万元的吴仲良教育奖励基金和敬老基金，受资助者近万人。吴仲良先生作为资阳第一代走出国门的川菜厨师，将川菜带向了世界，事业成功之后又以爱祖国、爱故乡、爱教育之心将自己的财富回报给祖国和故乡，值得人们铭记和敬仰。

In 1982, Wu Zhongliang went back to his hometown Ziyang to sweep the tombs. On seeing the humble rural schools and students' dropping out of school, he was hardly at peace and determined to make a contribution to the cause of education in his hometown. Since 1986, he donated his savings and part of his pensions amounted to over 20 million yuan to his hometown to initiate education business, built dozens of Qiaoxin schools, one kindergarten and one seniors' home successively in his hometown Lezhi County and Ziyang City, Neijiang City and many places, and set up Wu Zhongliang Education Award Fund and Fund for the Aged with an amount of more than 3.5 million yuan, with nearly ten thousand recipients. As the first generation of Sichuan cuisine chef in Ziyang who went abroad, Mr. Wu Zhongliang, brought Sichuan cuisine to the world. After his successful career, he returned his wealth to his motherland and hometown with the love for the motherland, hometown and education, which was worthy of people's memory and admiration.

5 CHEN SONGRU AND SICHUAN RESTAURANT 陈松如与四川饭店

陈松如（1921年~1993年），资阳老君场下坪村人，国家特一级烹调师，曾任北京四川饭店首席厨师长、技术总监和高级烹饪技师。陈松如是资阳籍的国宝级川菜烹饪大师，技艺精湛，曾多次为毛泽东、邓小平等党和国家领导人烹制菜肴，主理过许多重要的国宴，也代表中国赴世界各国展示、推广中华美食。陈松如大师善于理论总结、乐于传授技艺，撰写出版了多部川菜烹饪经典著作，培养了一批川菜名厨，在川菜烹饪发展史上留下了不可磨灭的功勋。

Chen Songru (1921—1993), a native of Xiaping Village, Laojun Field, Ziyang, was a national super-class I chef, who once served as Chief Chef, Technical Director and Senior Cooking Technician of Sichuan Restaurant in Beijing. Chen Songru was a national treasure of Sichuan cuisine master whose native place is Ziyang with exquisite skills. He cooked for Mao Zedong, Deng Xiaoping and other Party and state leaders for many times, presided over many important state banquets, and also exhibited and promoted Chinese cuisine on behalf of China to the world. Master Chen Songru was good at theoretical summary and willing to teach skills. He wrote and published many classic works of Sichuan cuisine, cultivated a number of famous chefs of Sichuan cuisine and left indelible achievements in the development history of Sichuan cuisine.

据陈松如手写的回忆录记载，1921年12月，他出生在资阳的一个贫苦农民家庭。为生活所迫，1933年，年仅12岁时就到成都，先后在民生饭店和成都饭店当学徒，后来又在多家餐馆事厨，1946年后进入当时在成都市饮食行业中规模较大的朵颐食堂，直到1957年。由于他博采众长，潜心钻研，技艺不断提高，主理、操办了许多大型会议的接待用餐，包括20余次且每次上千人的抗美援朝志愿军官兵用餐。

1957年，陈松如调往成都市圣灯寺国营餐厅任主任和厨师长。1959年，当时的北京正在组建四川饭店，陈松如奉命调入北京四川饭店担任厨师长，在四川饭店开业试餐后，受到前来参加试餐的周恩来、朱德、陈毅、邓小平等党和国家领导人的高度赞扬。陈松如还曾三进中南海，出色地完成了为毛泽东主席展示川菜技艺的任务。由此在很长一个时期，党和国家领导人宴请中外宾客或举行重大国事活动时都指定四川饭店的川菜为国宴的必上菜。1963年的一次国宴上，朝鲜的崔庸健委员长吃了陈松如做的家常海参后，对周恩来说："我们朝鲜也出海参，但就是做不出这样的味，我要派人来学习。" 1982年，邓小平在四川饭店为西哈努克亲王设宴，西哈努克品尝后说："今天的菜很好，我吃得肚子都圆了。" 邓小平非常高兴，专门与陈松如等厨师合影留念。陈松如大师在继承川菜"一菜一格，百菜百味"的基础上，本着博采众长、融会贯通、以我为主、自成一体的原则，结合北方气候和饮食习惯，按照国宴的规格和要求，创造了具有独特风格的四川饭店川菜体系，其中的数十个品种成为国宴的保留菜品。1986年在北京首批特级技师命名大会上，陈松如被授予"特一级中餐厨师"称号。1987年，他率6名弟子赴新加坡进行为期20天的烹饪表演，引起轰动，被当地报刊誉为中国国宝级川菜烹饪大师。1992年，他荣获世界名厨协会授予的"世界名厨"称号。

  According to his handwritten memoir, Chen Songru was born in a poor peasant family of Ziyang in December, 1921. Forced by life, he came to Chengdu in 1933 when he was only 12 years old, and served as an apprentice in Minsheng Restaurant and Chengdu Restaurant successively, and then worked as a cook in several restaurants. After 1946, he entered Duoyi Canteen, which was relatively large in the catering industry in Chengdu, and worked there until 1957. Because he learned widely from others, devoted himself to the study, and continuously promoted the skills, he presided and made arrangements of the receptions and meals of many large conferences, including more than 20 dinings of volunteers to the war to resist U.S. aggression and aid Korea, with each one involving thousands of

people. In 1957, Chen Songru was transferred to Chengdu Shengdengsi Temple State-Owned Restaurant as director and head chef. In 1959, when Beijing was setting up Sichuan Restaurant, Chen Songru was ordered to transfer to Sichuan Restaurant in Beijing as head chef. After the trial opening of Sichuan Restaurant, he was highly praised by Zhou Enlai, Zhu De, Chen Yi, Deng Xiaoping and other Party and state leaders who came for the trial meal. Chen Songru also entered Zhongnanhai three times and excellently completed the task of demonstrating Sichuan cuisine skills for Chairman Mao Zedong. Therefore, for a long time, party and state leaders designated Sichuan cuisine in Sichuan Restaurant as the essential dishes for state banquets when entertaining Chinese and foreign guests or holding major state events. At a state dinner in 1963, Chairman Cui Yongjian of the Democratic People's Republic of Korea said to Zhou Enlai after eating the Home Style Sea Cucumber cooked by Chen Songru, "We also produce sea cucumber, but we can't make such a taste. I want to send someone to study it." In 1982, Deng Xiaoping gave a banquet for Prince Norodom Sihanouk at the Sichuan Restaurant. Sihanouk tasted the food and said, "Today's food is very good. I've got a round stomach." Deng Xiaoping was very happy and specially took a group photo with Chen Songru and other chefs. On the basis of inheriting Sichuan cuisine of "one dish one style, one hundred dishes and one hundred flavors", in line with the principles of learning widely from others, integrating a thorough knowledge, placing emphasis on self, having a unique style, as well as combining northern climate and eating habits, and according to the specifications and requirements of the state banquet, Master Chen Songru created a unique style of Sichuan cuisine system of the Sichuan Restaurant, among which dozens of varieties have become reserved dishes for state banquets. In 1986, Chen Songru was awarded the title of "Super-Class I Chinese Cuisine Chef" at Beijing First Batch of Super Technicians Naming Conference. In 1987, he led six disciples to Singapore for a 20-day cooking performance, which caused a sensation, and he was praised as China' national treasure of Sichuan cuisine master by local newspapers. In 1992, he was awarded the title of "World Famous Chef" by World Famous Chef Association.

陈松如大师不仅技艺精湛，在烹饪技术理论上也颇有造诣。他利用业余时间精心编写出版了《北京四川饭店菜谱》《正宗川菜160例》，并在国内外数十种报刊及多家电台、电视台阐述菜点制作技艺。其中，《正宗川菜160种》一书内容丰富，文字简洁，通俗易懂，不仅介绍了川菜的历史发展和特色，还阐述了正宗川菜、小吃的制作方法，每一例都附有彩色照片，是烹制正宗川菜的权威性读物。此外，半个多世纪以来，他还培养了一大批川菜人才，其弟子遍布全国各地，许多人已成为川菜名厨，为川菜人才队伍建设做出了突出贡献。

Master Chen Songru was not only skilled, but also accomplished in culinary theory. In his spare time, he meticulously compiled and published *Recipes of Sichuan Restaurant of Beijing* and *160 Dishes of Authentic Sichuan Cuisine*, and elaborated on the cooking techniques of dishes in dozens of newspapers and magazines at home and abroad, as well as in many radio and television stations. Among them, the book *160 Dishes of Authentic Sichuan Cuisine* is rich in content, concise in text and easy to understand. It not only introduces the historical development and characteristics of Sichuan cuisine, but also expounds the making methods of authentic Sichuan cuisine and snacks. With each dish being accompanied with color photographs, it is the authoritative reading material for cooking authentic Sichuan cuisine. In addition, for over half a century, he also cultivated a large number of talents for Sichuan cuisine. His disciples are all over the country, and many people have become famous chefs of Sichuan cuisine, which has made outstanding contributions for the construction of Sichuan cuisine talent team.

6 肖述明与柠檬风味宴 XIAO SHUMING AND LEMON FLAVOR BANQUET

肖述明（1965年—），资阳市安岳县岳阳镇人，"柠檬风味宴"创始人，中国烹饪大师，曾荣获中国餐饮辉煌30年全国优秀工作者、川菜辉煌30年匠心功勋人物奖和四川省川菜全奖技术能手、资阳市高技能杰出人才等称号。

Xiao Shuming (1965—), born in Yueyang Town, Anyue County, Ziyang City, is the founder of the "Lemon Flavor Banquet" and the China Cooking Master, who has won the titles of National Excellent Worker of Chinese Catering Brilliant 30 Years, Sichuan Cuisine Brilliant 30 Years Master of Craftsmanship Award, Sichuan Cuisine Full Award Technical Expert of Sichuan Province, High-Skilled and Outstanding Talent of Ziyang City, etc.

肖述明出生于烹饪世家，从小就对烹饪产生了浓厚兴趣，13岁时便开始学厨，一边在饭店跟随师父学艺，一边勤奋钻研父亲留下的四川饭店老菜谱，烹饪技艺不断提高。经过20余年的打拼，他拥有了自己的酒店，便开始思考如何将安岳特产柠檬与川菜结合，融入川菜之中，由此开始了柠檬系列菜、柠檬风味宴的创制历程。肖述明带领研发团队，经过努力学习、刻苦钻研，充分利用柠檬的酸鲜爽朗，以安岳柠檬作为主料、辅料、调料，研发、制作出柠檬菜肴50余个，创制出"柠檬风味宴"，并于2009年10月成功申报为"中国·四川名宴"，2018年被中国烹饪协会评为"中国地域菜系四川十大主题名宴"之一。

Born in a cooking family, Xiao Shuming had a strong interest in cooking since childhood, and began to learn to cook when he was 13 years old. While learning from his master in the restaurant, he diligently studied the old Sichuan restaurant recipes left by his father. Thus the cooking skills continued to improve. After more than 20 years of struggle, he owned his own hotel, began to think about how to combine lemon, the Anyue specialty, with Sichuan cuisine, and integrate it into Sichuan cuisine, and thus began the creation process of Series of Lemon Dishes, and Lemon Flavor Banquet. Led by Xiao Shuming, the research and development team developed and produced more than 50 lemon dishes with lemons as main material, ingredient, or seasoning by making full use of the acid and the refreshment of lemons through hard study, and thus created the "Lemon Flavor Banquet", which was successfully declared as "China · Sichuan famous Banquet" in October, 2009, and was rated as one of the "Chinese Regional Cuisine Sichuan Top Ten Theme Banquets" by China Cuisine Association in 2018.

为推广柠檬系列菜及柠檬风味宴，肖述明先后到南京、衡阳、重庆、成都、雅安、内江、遂宁、大邑、宜宾等地进行宣传和展示，组织人员参加各级各类烹饪大赛，还参与编写《美食安岳》一书。中央电视台、四川电视台和重庆市、资阳市、安岳县等电视台，以及省内外多家杂志都对柠檬宴及其饮食文化进行宣传报道。如今，柠檬美食文化已成为"中国柠檬之乡"安岳的一张靓丽名片，肖述明不仅是"柠檬风味宴"的创始人，还培养了一批川菜烹饪人才，促进了资阳美食的创新发展。

In order to promote the series of lemon dishes and Lemon Flavor Banquet, Xiao Shuming successively went to Nanjing, Hengyang, Chongqing, Chengdu, Ya'an, Neijiang, Suining, Dayi, Yibin and other cities for publicity and display, organized personnel to attend all kinds of cooking competitions, and also participated in the compilation of the book *Gourmet Anyue*. CCTV, Sichuan TV, and TV stations of Chongqing, Ziyang, Anyue and other cities, as well as a number of magazines inside and outside Sichuan were all publicized and reported on the Lemon Flavor Banquet and its food culture. Nowadays, lemon food culture has already become a beautiful name card of the "Hometown of Lemon in China", Anyue. Xiao Shuming not only initiated the "Lemon Flavor Banquet", but also cultivated a batch of talents for Sichuan cuisine cooking, and promoted the innovative development of Ziyang cuisine.

7 LIU GUANYIN AND CHANGHONG FOOD CULTURE BANQUET 刘官银与苌弘美食文化宴

刘官银（1964年—），资阳市雁江区人，资阳市餐饮协会会长，资阳市金迪大酒店董事长，被评为四川省劳动模范、四川省杰出民营企业家和资阳市三贤文化传承使者。他积极搭建政府与餐饮企业之间的桥梁，坚持创新理念，倡导诚信经营，践行社会责任，大力推动资阳特色餐饮品牌打造和产业发展。他带领研发团队，深入挖掘资阳地域美食与历史人文资源，经过多年不懈努力研究创制的"苌弘美食文化宴"被评为四川名宴，逐渐成为资阳市对外交往的城市名片之一。

Liu Guanyin (1964—), born in Yanjiang District of Ziyang City, is the President of Ziyang Catering Association and Chairman of Ziyang Jindi Hotel, who has been rated as the Model Worker of Sichuan Province, Outstanding Private Entrepreneur of Sichuan Province and Inheritor of Three Sages Culture in Ziyang. He actively builds a bridge between the government and the catering enterprises, insists on innovative ideas, advocates honest management, practices social responsibility, and vigorously promotes the construction of Ziyang characteristic catering brands and the industrial development. He led the research and development team to deeply explore the regional food and historical and cultural resources of Ziyang. After years of unremitting efforts, research and creation, "Changhong Food Culture Banquet" has been rated as the Famous Banquet in Sichuan, and has gradually become one of Ziyang's name cards for foreign exchanges.

刘官银认为，消费需求始终是引导产业发展的航标，应深入挖掘本地特色食材和文化资源，积极打造资阳特色美食品牌，弘扬美食文化。2006年，刘官银作为资阳市民营企业家，在打造资阳三贤文化主题酒店——金迪大酒店的过程中，聘请秦照明作为顾问，与厨师们一起成立研究小组，通过多方挖掘与调研，收集到资阳三贤之一苌弘所著《苌膳斋》的口传膳谱，并以此为基础，结合苌弘的人生经历与资阳特色食材，精心研发、创制出"苌弘养生文化宴"。该宴的部分经典名菜有苌弘鲶鱼、圣贤思乡肘、碧血丹心、苌弘智多星、弘搏中原、舟游四海、高岩仙境、廖家莲藕、长生贡桔等。经不断提升和完善，苌弘美食文化宴被评为四川名宴。2011年，苌弘美食文化宴代表资阳市参加四川省首届地方旅游特色菜大赛，在全省31支代表队中荣获团体金奖、最佳营养配膳奖、最佳创意奖、最佳色彩奖等奖项，取得了7金、2银的优异成绩。如今，苌弘美食文化宴已成为资阳市对外交往的城市名片之一，刘官银也为资阳美食文化的传承、弘扬做出了应有的贡献。

Liu Guanyin believes that consumer demand is always the heading marker of guiding the development of the industry. It is necessary to dig deep into the local characteristic food materials and cultural resources, actively build Ziyang characteristic food brands and carry forward food culture. As a private entrepreneur of Ziyang, Liu Guanyin invited Qin Zhaoming as the consultant and set up the research team together with the chefs in the process of building Ziyang Sanxin Cultural Theme Hotel—Jindi Hotel in 2006. Via multiple research and investigation, they collected the oral recipes of *Chang Shan Zhai* written by Chang Hong, who is one of the Three Sages of Ziyang. On this basis, and combining with Chang Hong's life experience and Ziyang special food materials, they carefully researched, developed and created the "Changhong Health Care Culture Banquet". Some classic dishes in this banquet include Changhong Catfish, Sage Homesick Elbow, Blue Blood and Red Heart, Changhong Mastermind, Hong Bo Central Plains, Shipping the Four Seas, Fairyland of High Rocks, Liao's Lotus Root, Orange for Longevity and so on. Through constant improvement and perfection, Changhong Food Culture Banquet has been rated as the Famous Banquet in Sichuan. In 2011, Changhong Food Culture Banquet participated in The First Local Tourism Specialty Competition of Sichuan Province on behalf of Ziyang City, and won Group Gold Award, Best Nutrition Catering Award, Best Creative Award, Best Color Award and so on among 31 teams of the whole province, and won 7 gold and 2 silver awards. Today, Changhong Food Culture Banquet has become one of Ziyang's name cards for foreign exchanges, and Liu Guanyin has made due contributions to the inheritance and promotion of the food culture in this city.

许翔与桑叶系列菜 XU XIANG AND SERIES OF MULBERRY LEAF DISHES 8

许翔（1964年—），资阳市乐至县人，乐至县小贝壳酒店董事长，荣获资阳市餐饮业发展特别贡献奖。他长期致力于川菜经营与乐至民间美食的传承与创新，利用乐至蚕桑产业大县的优势，研发、创制出桑叶系列菜。

Xu Xiang (1964—), born in Lezhi County of Ziyang City, Chairman of Little Shell Hotel of Lezhi County, has once won the Special Contribution Award for Catering Industry Development of Ziyang City. He has long been committed to the management of Sichuan cuisine and the inheritance and innovation of Lezhi folk cuisine, taking advantages of Lezhi sericulture industry to develop and create the series of mulberry leaf dishes.

许翔从小热爱烹饪，勤于思考，家乡丰富的美食文化资源为他的餐饮之路打下了坚实基础。1993年，许翔赴西安开办酒楼、经营川菜，1996年回到乐至创建小贝壳酒楼。该酒楼的名称源于儿时在家乡河边拾贝的回忆，寄托着他创建一个美好、真诚的美食文化空间的追求。经过20余年的努力，许翔的酒楼已从不到300平方米的川菜小馆发展到营业面积达10 000平方米的资阳市一家著名餐饮企业。多年来，许翔一直坚持"产品与服务是生命线"的理念，带领团队努力挖掘家乡美食精华，并加以改良、创新，走出了一条乐至县餐饮地方特色化创新之路。他不仅深度挖掘书法家谢无量回乐至老家，在回澜镇品尝并赞赏红烧肉的轶事，创制出"无量醪糟红烧肉"等名菜，更利用乐至蚕桑产业大县的优势，以食用桑叶为食材，精心研发、创制了桑叶系列菜，包括桑叶酥、桑叶薄脆、桑汁浸白肉、倒罐蒸桑叶土鸡等菜品。如桑汁浸白肉，是在传统川菜蒜泥白肉的基础上演变而来，因冬季气温低，白肉易凝结、口感肥腻，便以温热的桑叶汁浸泡，既能增香、去腻，也能使肥肉部分呈半透明状，增添食欲。如今，经他牵头创制的桑叶系列菜和地方美食受到食客广泛称赞，助推了乐至县桑城品牌的打造。

Xu Xiang has loved cooking since childhood and is diligent in thinking. The rich food culture resources in his hometown have laid a solid foundation for his catering career. In 1993, Xu Xiang went to Xi'an to open restaurants and operate Sichuan cuisine, and got back to Lezhi in 1996 to set up Little Shell Hotel. The name of the restaurant originates from his childhood memories of picking up shellfish by the river in his hometown, which is the sustenance of his pursuit of creating a fine and sincere food culture space. After more than 20 years of efforts, Xu Xiang's restaurant has grown from a small Sichuan cuisine restaurant of less than 300 square meters into a famous catering enterprise in Ziyang City with a business area of 10,000 square meters. Over the years, Xu Xiang has been adhering to the concept of "products and services are the lifeline", leading the team to explore the essence of hometown dishes, and to make improvement and innovation, and walking out of a road of local characteristics innovation of Lezhi catering. He not only deeply dug the anecdote of calligrapher Xie Wuliang going back to his hometown Lezhi and tasting and praising the Braised Pork in Brown Sauce in Huilan Town to create "Wuliang Braised Pork in Brown Sauce with Fermented Glutinous Rice" and other famous dishes, but also took advantages of Lezhi sericulture industry with edible mulberry leaves as food materials to exquisitely develop and create the series of mulberry leaf dishes, such as Mulberry Leaf Pastry, Mulberry Leaf Pancake, Sliced Pork in Mulberry Juice, Poured Pot Steamed Mulberry Leaf Local Chicken, etc. Taking the Sliced Pork in Mulberry Juice as an example, it is created on the basis of the traditional Sichuan cuisine Sliced Pork with Garlic Sauce. Because of the low temperature in winter, the pork slices are easy to stick together and have greasy taste. Soaking the pork slices in warm mulberry leaf juice can increase incense and remove greasy flavor; moreover, it can make the fat part translucent and add appetite. Today, the series of mulberry leaf dishes created by the team led by him and the local cuisine are widely praised by diners, helping build the Sang Cheng brand of Lezhi County.

9 QIN ZHAOMING AND ZIYANG FOOD CULTURAL AND CREATIVE INDUSTRIES 秦照明与资阳美食文创

秦照明（1953年—），资阳市人，四川省烹饪协会副秘书长，资阳市餐饮协会执行副会长兼秘书长。他不仅积极开展餐饮协会的各项工作，搭建政府与餐饮企业之间的桥梁，而且挖掘、传承和弘扬资阳美食文化，是资阳美食文创的重要推动者和实践者。

Qin Zhaoming (1953—), born in Ziyang, is the Deputy Secretary General of Sichuan Cuisine Association, the Executive Vice President and Secretary General of Ziyang Restaurant and Catering Association. He not only actively carries out the work of restaurant and catering association, builds the bridge between the government and the catering enterprises, but also excavates, inherits and carries forward Ziyang food culture. He is an important promoter and practitioners of Ziyang food cultural and creative industries.

秦照明擅长餐饮经营管理，大力培养餐饮人才，也十分注重美食文化创意。自1986年以来，先后选送200余人到四川旅游学院、四川省旅游学校和四川省烹饪协会学习与培训，如今有17人在全国三星级酒店以上担任总经理，60余人成为资阳市许多酒店和餐饮企业的总经理、副总经理、部门经理，还为资阳市推荐选培了中国烹饪大师、名师8人。进入21世纪以后，随着经济发展和人们追求美好幸福生活的要求越来越高，餐饮酒店服务产业逐渐受到各级政府以及社会的重视，各种相关的鼓励和优惠政策相继出台。在此背景下，他率先向金迪大酒店董事长刘官银提出结合资阳的历史文化、地域文化、名人文化、民风民俗文化、风景区文化进行酒店餐饮创新发展的理念，实行文化兴企、创建品牌，并作为顾问，参与了"苌弘美食文化宴"研发，并获得成功。此后，他又先后指导多家酒店创意、研发了"同福家园美食文化宴""乐至桑都美食文化宴"等资阳地域文化主题宴。此外,他还组织将苌弘鲶鱼、外婆坛子肉传统制作技艺等申报并获批列入资阳市非物质文化遗产代表性名录。这些举措都树立了资阳美食文化创意的品牌效应,为讲好资阳餐饮故事、做好"资味"特色菜品、创建"资味"品牌贡献了自己的力量。

Qin Zhaoming is good at catering management, vigorously cultivates catering talents, and also attaches great importance to food culture and creativity. Since 1986, more than 200 people have been selected to study and train in Sichuan Tourism University, Sichuan Institute of Tourism and Sichuan Cuisine Association. Now, 17 people are working as general managers in national three-star and above hotels and more than 60 people have become general managers, deputy general managers and department managers of many hotels and catering enterprises in Ziyang. He also recommended and selected 8 Chinese cooking masters and top cooking teachers for Ziyang. After entering the 21st century, with the development of economy and the increasing requirements of people's pursuit of a better and happier life, the catering and hotel service industry has gradually received the attention of government and society at all levels, and various relevant encouragement and preferential policies have been issued one after another. Under this background, he took the lead to put forward to Liu Guanyin, the Chairman of Jindi Hotel, the idea of combining innovative development of hotel catering with historic culture, regional culture, celebrity culture, folk customs culture and scenic area culture, and implemented the strategies of culture prospering enterprises and creating brands. And as a consultant, he participated in the research and development of "Changhong Food Culture Banquet", which was very successful. Since then, he has guided a number of hotels to create, research and develop the "Tongfu Homeland Food Culture Banquet", "Lezhi Mulberry Capital Food Culture Banquet" and other regional cultural theme banquets of Ziyang. In addition, he organized the declarations of traditional production techniques of Changhong Catfish and Diced Pork in Grandma's Pot and were approved to be included in the representative list of Ziyang municipal intangible cultural heritage. All these measures have set up the brand effect of Ziyang's food culture and creativity, contributing to telling a good story of Ziyang catering, making "Zi Wei" special dishes and creating "Zi Wei" brand.

Chapter Six

ZIYANG FAMOUS SPECIAL LIQUORS AND TEA DRINKS

Ziyang has abundant water resources and natural resources, which provides a good material basis for local liquor brewing and drink manufacturing. Ziyang has a long history of liquor brewing. It was very popular in the Song Dynasty. According to the records of *Ziyang Annals* and other historical materials, the local liquor tax collected during the Xining Period of the Song Dynasty reached "more than 30,000 guan" (Guan is a currency unit of ancient China). In the tenth year of Song Shaoxing Period (1140 A. D.), there were officials who "supervised the liquor tax in Ziyang County". During Ming and Qing Dynasties, the "Wushi Dry Liquor" of Ziyang gradually became famous throughout Sichuan. Nowadays, Ziyang has a large number of special, famous and excellent liquors and tea drinks, among which the wines and tea drinks made from lemon, mulberry leaves and mulberries are distinctive and unique in flavor.

第六篇 资阳

名特酒品与茶饮

资阳市的水资源丰富，物产繁多，为当地酿酒和饮品提供了良好的物质基础。资阳酿酒历史悠久，到宋代时已十分盛行。据《资阳县志》等史料记载，宋代熙宁年间，当地征收的酒课即酒税就达「三万贯以上」（「贯」是中国古代的一种货币单位），宋绍兴十年（公元1140年）设有「监资阳县酒税」的官吏。明清时期，资阳酿制的「伍市干酒」逐渐闻名全川。如今，资阳拥有众多的名特优质酒品和茶饮，其中以柠檬、桑叶与桑葚制作的茶酒品种更是独树一帜、别具风味。

FAMOUS SPECIAL LIQUORS
名特酒品

宝莲大曲　BAOLIAN DAQU　1

宝莲大曲是产自资阳的特产名酒,由明朝永乐年间"伍市干酒"演变而来,其酿造工艺及配料秘方历经600余年的传承、精进,因其"绵柔淡雅,纯正甘爽"的独特风格而长期受到消费者青睐,在中国浓香型白酒中独树一帜。宝莲大曲酒液无色透明,无悬浮物,窖香浓郁,醇和爽净,诸味协调,余香悠长,后味干净,酒度有38°、54°两种。

Baolian Daqu is famous specialty liquor produced in Ziyang, which was evolved from "Wushi Dry Liquor" of Yongle Period of Ming Dynasty. After more than 600 years of inheritance and refinement of its brewing technique and secret ingredients, it has long been favored by consumers due to its unique style of "being soft and elegant, pure and refreshing", and has developed a school of its own in Chinese Luzhou-flavor liquor. The liquor of Baolian Daqu is colorless and transparent, has no floating substances, with strong cellar aroma. It tastes mellow and clean, all flavor coordinated, with lingering aroma and clean after-taste. There are two Baolian Daqu degrees of 38° and 54°.

宝莲大曲历史悠久。资阳在宋代已盛行酿酒。据《资阳县志》载,宋绍兴十年(公元1140年)设有"监资阳县酒税"官吏。清代康熙年间酿有"资阳陈色,伍市干酒","煮酒贩运成都",有"过龙泉香十里"之称。1951年,在旧作坊基础上成立四川省资阳县国营第十八酒厂，酿制伍市干酒。宝莲大曲选用优质高粱、小麦为原料，加入优质清水，以小麦特制成的包包曲为糖化发酵剂，采用固态续糟混蒸工艺和多种工序精心酿造而成。宝莲大曲先后荣获四川省优质产品和商业部优质产品、国家优质酒等称号。

Baolian Daqu has a long history. Liquor brewing was prevalent in Ziyang in the Song Dynasty. According to the records of *Ziyang Annals*, in the tenth year of Song Shaoxing Period (1140 A. D.), there were officials who "supervised the liquor tax in Ziyang County". During Kangxi Period of Qing Dynasty, "Ziyang Chense and Wushi Dry Liquor" were brewed, "cooking and trading in Chengdu", which were well known for "lingering aroma of ten li beyond Longquan". In 1951, Sichuan Ziyang State-owned 18th Liquor Factory was established on the basis of the old workshop to make Wushi Dry Liquor. Baolian Daqu is brewed carefully by using high quality sorghum and wheat as raw materials, adding high quality water, using special wheat wrapped starter as saccharification leavening agent, and adopting solid state continuous distilling technique and various other processes. Baolian Daqu has successively obtained the titles of Quality Products of Sichuan Province, Quality Products of Ministry of Commerce, National Quality Liquor, etc.

2 LIANGJIESHAN AGED LIQUOR 两节山老酒

两节山老酒因产自资阳两节山而得名，也同样源自永乐年间的"伍市干酒"。两节山老酒传统酿造技艺于2013年被资阳市人民政府列入市级非物质文化遗产代表性名录。

Liangjieshan Aged Liquor gets its name from Liangjieshan mountain of Ziyang. It also originated from the "Wushi Dry Liquor" which was originated in Yongle Period. In 2013, the traditional brewing technique of Liangjieshan Aged Liquor was included in the representative list of municipal intangible cultural heritage by Ziyang Municipal People's Government.

两节山老酒以当地小糯红高粱为主要原料，加入两节山一带无色透明、清香回甜的优质井泉水及酒曲，采用独特的传统清蒸续糟工艺和多种工序，以数百年来传承使用的香樟木甑子装酿。此酒在两节山独特的地理气候和酿酒微生物环境酿造而成，具有口感纯正、窖香浓郁、醇甜爽净、余味悠长、挂杯留香等特征，深受历代老百姓的喜爱，享有近六百年的盛誉。

Liangjieshan Aged Liquor takes local small glutinous red sorghum as the main raw material, adds starter and high quality Liangjieshan well and spring water of colorless and transparent, fragrant and sweet, adopts unique traditional steamed continuous distilling technique and various processes, and brews in the camphorwood Zengzi inherited and used for hundreds of years. The liquor is brewed in the unique geographical climate and microbial environment for brewing of Liangjieshan Mountain, which has the characteristics of pure taste, strong cellar aroma, mellow sweet and clean, long aftertaste, tears of glass, and lingering aroma. It enjoys a high reputation of nearly 600 years and is deeply loved by people.

两节山老酒始创于明代，烤酒匠人后裔黄国仲在雁江区石岭镇培德场开办"浸水阁烧坊"，烤出的白酒纯正柔和、余味悠长，远近闻名，到清代康熙年间，其知名度更是不断提升，以致享誉全川。相传康熙年间，20余名挑夫挑着老酒去成都，不料在龙泉山顶窝窝店突遇成都府的八抬大轿，慌乱中打碎了酒坛。成都知府被酒香陶醉，非但没有责怪治罪于轿夫，反而让书吏记下酒名。此后两百多年，此酒成为成都府衙门里的"官酒"。清代末年以后被成都各大酒庄竞相抢购。当时的知名报纸《新新新闻》称赞："无色透明，光鉴毫芒。异香四溢，沁人脾肠。醇浓味美，口齿留芳。" 1942年《资阳县志》记载了两节山老酒的由来，20世纪50年代以后，两节山老酒经过技术革新和工艺改良，多次被评为优质酒。两节山老酒酿造技艺薪火相传，成为资阳市知名的白酒品种。

　　Liangjieshan Aged Liquor was created in the Ming Dynasty. Huang Guozhong, a descendant of the roasting liquor craftsman, opened the "Jinshuige Liquor Workshop" in Peide Field, Shiling Town, Yanjiang District. The liquor roasted by the workshop was known far and wide for it was pure and soft, with a long aftertaste. By Kangxi Period of the Qing Dynasty, its popularity increasingly improved and it became famous all over Sichuan. It is said that during the Kangxi Period, more than 20 men carried the aged liquor to Chengdu. Unexpectedly, they suddenly encountered the large sedan chair of Chengdu Magistrate at the Wowo Shop on the top of Longquan Mountain, and broke the liquor in the confusion. Chengdu Magistrate was attracted by the aroma of the liquor. Instead of blaming the sedan bearers, he let the scribe write down the name of the liquor. More than 200 years since then, this liquor became the "Official Liquor" of Chengdu Government. After the end of the Qing Dynasty, it was snapped up by the major liquor manors in Chengdu. *Xinxin News*, a popular newspaper at the time, praised the liquor as "colorless and transparent, bright and ligh. Extraordinary fragrance wafts, refreshing spleen, with mellow flavor and fragrant taste." In 1942, *Ziyang Annals* recorded the origin of Liangjieshan Aged Liquor. After the 1950s, Liangjieshan Aged Liquor was rated as high quality liquor for many times after technical innovation and technology improvement. Liangjieshan Aged Liquor's brewing skills have passed down. It has become a famous liquor variety of Ziyang City.

外交家酒　DIPLOMAT LIQUOR　3

　　外交家酒因产自伟大的无产阶级革命家、军事家和外交家陈毅元帅的故里——资阳市乐至县而得名。陈毅元帅曾担任中华人民共和国外交部长，对我国的外交事业作出了巨大贡献。外交家酒旨在纪念卓越的外交家陈毅元帅，由原外交部部长姬鹏飞亲笔题书酒名："元帅之乡品美酒，外交酒香飘天下"。

　　Diplomat Liquor is named for it is produced in Lezhi County, Ziyang City—the hometown of Marshal Chen Yi, the great proletarian revolutionist, militarist and diplomat. Marshal Chen Yi served as Minister of Foreign Affairs of the People's Republic of China and made great contributions to the diplomatic cause. The liquor was made to commemorate the distinguished diplomat marshal Chen Yi, and was inscribed by Ji Pengfei, the former Minister of Foreign Affairs, with the words of "Tasting Mellow Liquor in Marshal's Hometown, Spreading Fragrance of Diplomat Liquor all over the World".

外交家酒起源于1911年的杂粮酒毛氏烧酒作坊,沿袭百年的酿造工艺。乐至县保留有百年老窖池48口,窖池中不仅具有多种天然微生物,更富含各种微量元素。良好的自然生态环境,使窖底丰富的有益生物群系赋予酒糟发酵而酯化生香。"千年老窖万年糟"是浓香型白酒主体香味的原载体,正是来源于窖底的窖香、糟香和五粮之粮香的完美结合,并经过时间的历练和沉淀,成就了醇厚香浓的佳酿。外交家酒先后获得国家商业部、四川省优质产品称号和乌兰巴托国际博览会金奖等多项大奖,在纪念陈毅元帅100周年庆祝活动中成为唯一指定专用白酒。

Diplomat Liquor originated from cereal liquor in 1911 in Mao's Liquor Workshop, and followed a century-old brewing technology. There are 48 a century-old cellar pools in Lezhi County, which not only has many natural microorganisms, but also is rich in various microelements. Good natural ecological environment makes the abundant beneficial biomes at the bottom of the cellar endow the fermentation of the distillers' grains with esterification and fragrance. "Thousands of years old cellar and ten thousands of years of distillers' grains" are the original carriers of the main fragrance of Luzhou-flavor liquor. It is the perfect combination and years of experience and precipitation of the aroma from the bottom of the cellar, distillers' grains aroma and five-grain aroma that make the mellow and fragrant liquor. Diplomat Liquor has successively obtained the titles of Quality Products of Ministry of Commerce and Quality Products of Sichuan Province; it also has won several awards, such as the gold medal of Ulan Bator International Expo. Besides, it is the only designated special liquor in Marshal Chen Yi's 100th Anniversary Celebration.

4 LEYI CELLAR LIQUOR 乐意窖酒

乐意窖酒是四川老字号名酒,产自资阳市乐至县。乐意窖藏酒制作技艺于2018年被资阳市人民政府列入市级非物质文化遗产代表性名录。

Leyi Cellar Liquor is Sichuan time-honored famous liquor, made in Lezhi County, Ziyang City. In 2018, the manufacture skill of Leyi Cellar Liquor was included in the representative list of municipal intangible cultural heritage by Ziyang Municipal People's Government.

乐意窖藏酒制作技艺先后经历了6代传人的传承,技艺更加精湛。乐意窖酒还借鉴泸州老窖和五粮液的相关技艺,以高粱、大米、糯米、小麦和玉米五种粮食为原料准确配比,汲取乐至县的优质泉水,老窖发酵、陶坛存储、精心勾调而成,具有窖香浓郁、绵软甘冽、醇厚协调、尾净余长的特点，深受消费者的喜爱，得到社会各界的赞誉。乐意窖酒曾作为国家科考队首赴南极考察指定专用酒。

The manufacture skill of Leyi Cellar Liquor has been passed down for 6 generations, which makes the skill more exquisite. Leyi Cellar Liquor also learns from the relevant skills of Luzhou Laojiao and Wuliangye. It is brewed by taking sorghum, rice, glutinous rice, wheat and maize as raw materials with accurate ratio, drawing high quality spring water of Lezhi, fermenting in aged cellar and storing in pottery jars, as well as carefully blending. It has the characteristics of strong cellar aroma, being soft and sweet, mellow coordination, and long aftertaste. Thus it is deeply loved by consumers, and gets great praise from all walks of life. Leyi Cellar Liquor was once served as the designated special liquor for the first expedition to Antarctica of the national scientific research team.

桑葚紫酒 | MULBERRY PURPLE WINE 5

桑葚紫酒是产自资阳市乐至县的特色优质果酒。乐至县具有悠久的桑蚕养殖历史，桑树种植量大，资源极为丰富。桑葚是桑树的果实，富含多种维生素、微量元素和蛋白质等，具有很高的营养价值，被称为"民间圣果"，被医学界誉为"21世纪的最佳保健果品"。乐至县出产的桑葚紫酒是发酵型桑葚果酒，经分选、压榨、果汁澄清、发酵、陈酿、冷处理、过滤、瞬时高温热杀、检验、自动装瓶、包装等工序，酒体果香浓郁，口感清爽鲜醇、柔和甜美，具有提高人体免疫力、强身抗老、美容养颜等作用。桑葚紫酒在中国第二十一届农洽会上获得优质产品奖，具有较高的知名度。

Mulberry Purple Wine is a special and high quality fruit wine produced in Lezhi County of Ziyang City. Lezhi County has a long history of silkworm cultivation. Mulberry trees are heavily planted, which provide abundant resources. Mulberry is the fruit of mulberry tree. It is rich in a variety of vitamins, microelements, proteins, etc. Besides, it has high nutritional value, and is known as "folk sage fruit", and is praised as "the best health care fruit in the 21st century" by the medical community. The Mulberry Purple Wine of Lezhi is a fermented mulberry fruit wine. After the processes of separation, crushing, juice clarification, fermentation, aging, cooling, filtering, transient high temperature thermal sterilization, inspection, automatic bottle filling and packing, the wine will be very fruity and intense and the taste will be fresh and mellow, soft and sweet. Moreover, it has the functions of improving human immunity, strengthening body and anti-aging, maintaining beauty, and so on. Mulberry Purple Wine is very popular and won the "High Quality Product Award" in The 21st China Agricultural Consultative Conference.

6　LEMON WINE　柠檬酒

柠檬酒是产自资阳市安岳县特色优质的发酵果酒，现已形成系列，主要包括柠檬果酒、柠檬蜜酒等品类。

Lemon Wine is a special and high-quality fermented fruit wine produced in Anyue County, Ziyang City. Now it has formed a series, mainly including lemon fruit wine, lemon honey wine and other categories.

安岳气候温和，雨量充沛，日照充足，无霜期长，适宜种植业发展。20世纪20年代，安岳引入美国的尤力克柠檬品种并广为种植，培育出了丰产、优质的柠檬新株系。安岳柠檬果实美观，品质上乘，是我国唯一的柠檬生产基地县，被命名为"中国柠檬之乡"。以安岳出产的柠檬为原料酿制的柠檬果酒，经过了前中后三期发酵、过滤、澄清、陈酿和包装灭菌等工序，酒体既有柠檬的清香又有酒特有的醇香，而且富含有机酸和维生素。如今，安岳柠檬果酒及柠檬蜜酒远销英国、阿拉伯联合酋长国、俄罗斯、新加坡、哈萨克斯坦等10余个国家及我国香港、澳门地区，在国内50余个大中城市也颇受欢迎。

Anyue has a mild climate, plenty of rain, abundant sunshine, long frost-free period, which is suitable for the development of planting industry. In the 1920s, Anyue introduced the American Eureka lemon and planted it widely, developing a new lemon strain with high yield and high quality. Anyue lemon has beautiful fruit and high quality. Anyue is China's only lemon production base county, and is named as "Hometown of Lemon in China". The lemon fruit wine made from Anyue lemons has gone through three stages of the process of fermentation, filtration, clarification, aging, packaging, sterilization, etc. The wine has a delicate smell of lemon and a special mellow aroma of liquor; it is rich in organic acids and vitamins. Nowadays, Anyue Lemon Fruit Wine and Lemon Honey Wine are exported to more than 10 countries including Britain, United Arab Emirates, Russia, Singapore, Kazakhstan, and so on. Also, they are popular in Hong Kong and Macao, as well as more than 50 large and medium-sized cities in China.

百威啤酒 | BUDWEISER BEER

百威啤酒诞生于1876年，是采用质量最佳的纯天然材料，以严谨的工艺控制，通过自然发酵、低温储藏而酿成的世界知名啤酒品牌。该啤酒在整个生产流程中不使用任何人造成分、添加剂或防腐剂。百威公司是世界知名的啤酒品牌企业和全球最大的啤酒生产、酿造商，位列世界500强。

Born in 1876, Budweiser Beer is a world-renowned beer brand brewed via natural fermentation and low temperature storage with the pure natural materials of best quality and strict technological control. The beer is produced without the use of any artificial ingredients, additives or preservatives. Budweiser is a world famous beer brand enterprise and the world's largest beer producer and brewer, ranking among the top 500 companies in the world.

2010年4月，百威公司在资阳正式签约成立百威（四川）啤酒有限公司，同时成立了百威（中国）销售有限公司资阳分公司，工厂占地面积593亩。资阳的百威啤酒注重从选料、糖化、发酵、过滤到罐装的每一个工序，以世界多个地区的优质麦芽、啤酒花、酵母等为原料，采用资阳地下100米的优质矿泉水和独特的工艺酿制而成，具有格外清澈、清爽、清醇的品质。从2011年投产至今，百威资阳工厂已生产啤酒超过230万吨，其产品已在四川、云南、贵州、西藏、新疆、山西、重庆等9省1市销售。

In April, 2010, Budweiser Company formally signed a contract to establish Budweiser (Sichuan) Beer Co., Ltd. in Ziyang, and set up the Ziyang Branch of Budweiser (China) Sales Co., Ltd., covering an area of 593 mu. Budweiser Beer in Ziyang focuses on every process from material selection, saccharization, fermentation, filtration to canning. It is brewed with high-quality malt, hops, yeast and other raw materials from many parts of the world, and with high-quality mineral water 100 meters underground in Ziyang and unique technology. Hence is has an exceptional quality of being clear, refreshing and mellow. Since its production in 2011, Budweiser Ziyang factory has produced more than 2.3 million tons of beer, and its products have been sold in nine provinces and one city, such as Sichuan, Yunnan, Guizhou, Tibet, Xinjiang, Shanxi, Chongqing and so on.

FAMOUS SPECIAL TEA DRINKS
名特茶饮

柠檬饮品　LEMON DRINKS　1

　　柠檬系列饮品是产自安岳县的特色优质饮料。安岳县作为"中国柠檬之都"，出产的柠檬含柠檬油7.4‰、可溶性固形物9.5%、柠檬酸6.7%、Vp2.5%、Vc58mg/100ml、果胶3%及丰富的微量元素等，出汁率38%，许多指标均超过美国、意大利柠檬，其品质堪称世界一流，是既能鲜食又能加工的最理想的优良品种。如今，安岳柠檬饮品的开发已成为完整的产业链，柠檬鲜果到了生产线，果皮用来生产柠檬蜜茶，果肉用来榨汁、制成柠檬饮料，果肉细渣还可制成柠檬蜜酱，剩下的柠檬籽用来制成花果茶。以柠檬为原料开发的柠檬茶（柠檬红茶、柠檬首乌茶）、蜂蜜柠檬、柠檬浓缩汁等系列饮品是安岳柠檬精深加工中的重要品种，深受消费者喜爱。在安岳，人们常常以一杯浓浓的柠檬汁佐餐，酸中带甜、清香爽口。

Lemon series of drinks are special and high quality drinks produced in Anyue. As the "Hometown of Lemon in China", the lemons produced there contain 7.4‰ lemon oil, 9.5% soluble solids, 6.7% citric acid, 2.5 % Vp, 58mg / 100 ml Vc, 3% pectin, rich microelements, and 38% juice yield. Many indicators exceed those of the lemons of the United States and Italy. It is of world-class quality and is the most ideal variety for fresh-eating and processing. At present, the development of Anyue lemon drinks has become a complete industrial chain. As fresh fruits arriving on the production line, the peel is used to produce lemon honey tea, the pulp is used to squeeze juice and make lemon drinks, the fine pomace can be made into lemon honey sauce, and the remaining lemon seeds are used to make flower fruit tea. Lemon Tea (Lemon Black Tea, Lemon Shouwu Tea), Honey Lemon, Lemon Concentrated Juice and other series of drinks developed with lemon as raw materials are important varieties in the deep-processing of Anyue lemon, which are deeply loved by consumers. In Anyue, people often eat with a cup of thick lemon juice, which is sour with sweet, pleasant to the palate.

2 MULBERRY BUD TEA 桑芽茶

桑芽茶是产自资阳市乐至县的一种优质茶饮品种，是用桑树嫩芽制作而成。乐至县以春季桑树的嫩芽为原料，通过杀青、干燥等加工工艺，制作出了色泽碧绿、风味清香、复水性好的桑芽茶。桑芽含有较高的黄酮化合物和桑苷等对人体有益的成分，具有一定的调节血压与血糖、润肠减肥、美容养颜和延缓衰老等作用。

Mulberry Bud Tea, which is made from mulberry bud, is a kind of high quality tea drink produced in Lezhi County, Ziyang City. Taking the shoots of mulberry trees in spring as the raw material, adopting fixation, drying and other processing technologies, Lezhi has produced the Mulberry Bud Tea which has bright green color, delicate flavor and good reconstitution properties. Mulberry bud contains high flavonoid compounds, mulberry glycosides, and other ingredients which are beneficial to human body. To some extent, it can regulate blood pressure and blood sugar, moisten intestines to lose weight, maintain beauty and keep young.

3 JDB HERBAL TEA 加多宝凉茶

凉茶是指用药性寒凉和能消解人体内热的中草药煎水、供人们饮用的饮料，最初常见于广东、香港和澳门等地，后来成为许多地区人们餐桌上的常用饮品。2006年，凉茶被批准列入第一批国家级非物质文化遗产代表性名录。

Herbal tea is a kind of decoction for people to drink, which is made from Chinese herbs that can dispel the inner heat of human body. It was originally common in Guangdong, Hong Kong and Macao, and later has become a

common drink on the dinner table in many areas. In 2006, herbal tea was approved to be included in the first batch of the representative list of national intangible cultural heritage.

加多宝凉茶是我国知名的罐装凉茶品牌。2014年，四川加多宝饮料有限公司的凉茶工厂在资阳雁江区正式建成投产。加多宝凉茶传承王泽邦先生始创于清朝道光年间的配方，秉承传统工艺，运用现代食品科技，有效提取本草植物精华，精心制作而成，内含菊花、甘草、仙草、金银花等清热解毒的植物。该工厂使用高速全自动罐装生产设备，每分钟可生产1 200罐凉茶，年产能1 800万箱。加多宝凉茶落户资阳市，为西部地区提供了一款口味纯正、品质卓越的凉茶饮品，满足了广大消费者的消费需求。

JDB Herbal Tea is a famous canned herbal tea brand in China. In 2014, the herbal tea factory of Sichuan JDB Beverage Co., Ltd. was officially built and put into production in Yanjiang District, Ziyang City. JDB herbal tea inherits the formula created by Mr. Wang Zebang during Daoguang Period of the Qing dynasty. It is elaborately made by adhering to the traditional technology, using modern food science and technology, effectively extracting the essence of herbs, including chrysanthemum, liquorice, mesona blume, honeysuckle and other heat-clearing and detoxifying plants. Using high-speed automatic canning production equipment, the factory can produce 1,200 cans of herbal tea per minute, with annual capacity of 18 million boxes. Being settled in Ziyang, JDB Herbal Tea provides western China with a herbal tea drink of pure taste and excellent quality, meeting the comsumption needs of the consumers.

資陽美食文化